APPLICATIONS OF ARTIFICIAL INTELLIGENCE IN BUSINESS AND FINANCE

Modern Trends

AAP Research Notes on Optimization and Decision-Making Theories

APPLICATIONS OF ARTIFICIAL INTELLIGENCE IN BUSINESS AND FINANCE

Modern Trends

Edited by

Vikas Garg, PhD

Shalini Aggarwal, PhD

Pooja Tiwari, PhD

Prasenjit Chatterjee, PhD

AAP | APPLE ACADEMIC PRESS

First edition published 2022

Apple Academic Press Inc.
1265 Goldenrod Circle, NE,
Palm Bay, FL 32905 USA
4164 Lakeshore Road, Burlington,
ON, L7L 1A4 Canada

CRC Press
6000 Broken Sound Parkway NW,
Suite 300, Boca Raton, FL 33487-2742 USA
2 Park Square, Milton Park,
Abingdon, Oxon, OX14 4RN UK

Library and Archives Canada Cataloguing in Publication

Title: Applications of artificial intelligence in business and finance : modern trends / edited by Vikas Garg, PhD, Shalini Aggarwal, PhD, Pooja Tiwari, PhD, Prasenjit Chatterjee, PhD.

Names: Garg, Vikas, editor. | Aggarwal, Shalini, editor. | Tiwari, Pooja, editor. | Chatterjee, Prasenjit, 1982- editor.

Description: First edition. | Series statement: AAP research notes on optimization and decision-making theories | Includes bibliographical references and index.

Identifiers: Canadiana (print) 20210242183 | Canadiana (ebook) 2021024223X | ISBN 9781771889681 (hardcover) | ISBN 9781774639658 (softcover) | ISBN 9781003129639 (ebook)

Subjects: LCSH: Financial services industry—Technological innovations. | LCSH: Business enterprises—Technological innovations. | LCSH: Artificial intelligence.

Classification: LCC HG173 .A67 2022 | DDC 332.10285—dc23

Library of Congress Cataloging-in-Publication Data

Names: Garg, Vikas, editor. | Aggarwal, Shalini, editor. | Tiwari, Pooja, editor | Chatterjee, Prasenjit.

Title: Applications of artificial intelligence in business and finance : modern trends / Vikas Garg, PhD, Shalini Aggarwal, PhD, Pooja Tiwari, PhD, Prasenjit Chatterjee, PhD.

Description: 1st edition. | Palm Bay, FL : Apple Academic Press, 2022. | Series: Aap research notes on optimization and decision making theories | Includes bibliographical references and index. | Summary: "In the current competitive and technology-driven world, AI has emerged as growing technology that is applicable to many fields. As transactions and other business functions move online and grow more popular every year, the finance and banking industries face increasingly complex data management and identity theft and fraud issues. AI can bring many financial and business functions to the next level, as systems using deep learning technologies are able to analyze patterns and spot suspicious behavior and potential fraud. In this volume, Applications of Artificial Intelligence in Business and Finance: Modern Trends, the focus is on the application of artificial intelligence in finance, business, and related areas. The book presents a selection of chapters presenting cutting-edge research on current business practices in finance and management. Topics cover the impact and application of AI on many diverse areas in finance and business, including e-commerce systems, financial services, fraud prevention, identifying loan-eligible customers, online business, Facebook social commerce, insurance industry, online marketing, and more. This book will be useful for those in the fields of finance and business management and also for professionals in electronic and communication technology. It will also be useful for academicians, postgraduate students, research associates, and others"-- Provided by publisher.

Identifiers: LCCN 2021028079 (print) | LCCN 2021028080 (ebook) | ISBN 9781771889681 (hardcover) | ISBN 9781774639658 (paperback) | ISBN 9781003129639 (ebook)

Subjects: LCSH: Business enterprises--Technological innovations. | Business--Data processing. | Finance--Data processing. | Artificial intelligence.

Classification: LCC HD45 .A687 2022 (print) | LCC HD45 (ebook) | DDC 658.5/14--dc23

LC record available at https://lccn.loc.gov/2021028079

LC ebook record available at https://lccn.loc.gov/2021028080

ISBN: 978-1-77188-968-1 (hbk)
ISBN: 978-1-77463-965-8 (pbk)
ISBN: 978-1-00312-963-9 (ebk)

AAP RESEARCH NOTES ON OPTIMIZATION AND DECISION-MAKING THEORIES

SERIES EDITORS:

Dr. Prasenjit Chatterjee
Department of Mechanical Engineering, MCKV Institute of Engineering, Howrah, West Bengal, India
E-Mail: dr.prasenjitchatterjee6@gmail.com / prasenjit2007@gmail.com

Dr. Dragan Pamucar
University of Defence, Military academy, Department of Logistics, Belgrade, Serbia; E-Mail: dpamucar@gmail.com

Dr. Morteza Yazdani
Department of Business & Management, Universidad Loyola Andalucia, Seville, Spain; E-Mail: morteza_yazdani21@yahoo.com

Dr. Anjali Awasthi
Associate Professor and Graduate Program Director (M.Eng.), Concordia Institute for Information Systems Engineering, Concordia, Canada
E-Mail: anjali.awasthi@concordia.ca

Most real-world search and optimization problems naturally involve multiple criteria as objectives. Different solutions may produce trade-offs (conflicting scenarios) among different objectives. A solution that is better with respect to one objective may be a compromising one for other objectives. This compels one to choose a solution that is optimal with respect to only one objective. Due to such constraints, multi-objective optimization problems (MOPs) are difficult to solve since the objectives usually conflict with each other. It is usually hard to find an optimal solution that satisfies all objectives from the mathematical point of view. In addition, it is quite common that the criteria of real-world MOPs encompass uncertain information, which becomes quite a challenging task for a decision maker to select the criteria. Also, the complexities involved in designing mathematical models increase. Considering, planning, and appropriate decision-making require the use of analytical methods that examine trade-offs; consider multiple scientific, political, economic, ecological, and social dimensions; and reduce possible conflicts in an optimizing framework. Among all these, real-world multi-criteria decision-making (MCDM) problems related to engineering optimizations are categorically important and are quite often encountered with a wide range of applicability.

MCDM problems are basically a fundamental issue in various fields, including applied mathematics, computer science, engineering, management, and operations research. MCDM models provide a useful way for modeling various real-world problems and are extensively used in many different types of systems, including, but not limited to, communications, mechanics, electronics, manufacturing, business management, logistics, supply chain, energy, urban development, waste management, and so forth.

In the aforementioned cases, modeling of multiple criteria problems often becomes more complex if the associated parameters are uncertain and imprecise in nature. Impreciseness or uncertainty exists within the parameters due to imperfect knowledge of information, measurement uncertainty, sampling uncertainty, mathematical modeling uncertainty, etc. Theories like probability theory, fuzzy set theory, type-2 fuzzy set theory, rough set, grey theory, neutrosophic uncertainty theory available in the existing literature deal with such uncertainties. Nevertheless, the uncertain multi-criteria characteristics in such problems are not explored in depth, and a lot can be achieved in this direction. Hence, different mathematical models of real-life multi-criteria optimization problems can be developed on various uncertain frameworks with special emphasis on sustainability, manufacturing, communications, biomedical, electronics, materials, energy, agriculture, environmental engineering, strategic management, flood risk management, supply chain, waste management, transportations, economics, and industrial engineering problems, to name a few.

Coverage & Approach:

The primary endeavor of this series is to introduce and explore contemporary research developments in a variety of rapidly growing decision-making areas. The volumes will deal with the following topics:

- Crisp MCDM models
- Rough set theory in MCDM
- Fuzzy MCDM
- Neutrosophic MCDM models
- Grey set theory
- Mathematical programming in MCDM
- Big data in MCDM

- Soft computing techniques
- Modelling in engineering applications
- Modeling in economic issues
- Waste management
- Agricultural practice
- Material selection
- Renewable energy planning
- Industry 4.0
- Sustainability
- Supply chain management
- Environmental policies
- Manufacturing processes planning
- Transportation and logistics
- Strategic management
- Natural resource management
- Biomedical applications
- Future studies and technology foresight
- MCDM in governance and planning
- MCDM and social issues
- MCDM in flood risk management
- New trends in multi-criteria evaluation
- Multi-criteria analysis in circular economy
- Multi-criteria evaluation for urban and regional planning
- Integrated MCDM approaches for modeling relevant applications and real-life problems

Types of volumes:

This series reports on current trends and advances in optimization and decision-making theories in a wider range of domains for academic and research institutes along with industrial organizations. The series will cover the following types of volumes:

- Authored volumes
- Edited volumes
- Conference proceedings
- Short research (thesis-based) books
- Monographs

Features of the volumes will include recent trends, model extensions, developments, real-time examples, case studies, and applications. The volumes aim to serve as valuable resources for undergraduate, postgraduate and doctoral students, as well as for researchers and professionals working in a wider range of areas.

CURRENT & FORTHCOMING BOOKS IN THE SERIES

Multi-Criteria Decision-Making Techniques in Waste Management: A Case Study of India
Editors: Suchismita Satapathy, Debesh Mishra, and Prasanjit Chatterje

Applications of Artificial Intelligence in Business and Finance: Modern Trends
Editors: Vikas Garg, Shalini Aggarwal, Pooja Tiwari, and Prasenjit Chatterjee

Computational Intelligence for Finance and Marketing in Post-Epidemic VUCA World
Editors: Subhranil Som, Ajay Rana, Prasenjit Chatterjee, Roshani Raut, and Hanaa Hachimi

Multi-Criteria Decision-Making Approaches for Designing Environmental Policies
Editors: Prasenjit Chatterjee, Dragan Pamucar, Morteza Yazdani, and Anjali Awasthi

Statistical Estimation Procedure under Model Based Optimization
Editors: Arnab Bandyopadhyay and Samir Dey

Intelligent Decision Support Systems for Smart City Application
Editors: Loveleen Gaur, Vernika Agarwal, and Prasenjit Chatterjee

ABOUT THE EDITORS

Vikas Garg, PhD

Director, Executive Programs Management Domain,
Amity University Uttar Pradesh, Greater Noida Campus, India

Vikas Garg, PhD, is the Director of Executive Programs Management Domain at Amity University Uttar Pradesh, Greater Noida Campus, India. He is UGC NET qualified with 15 years of academic experience. His areas of specialization are accounting and finance, and his primary interests are in financial markets, financial reporting, and analysis. He is a lifetime member of the Indian Commerce Association, Indian Accounting Association, and Indian Management Association. He is certified in Customer Relationship Management from the Indian Institute of Management Bangalore, India. He has published numerous research papers in various Scopus- and ABDC-indexed international and national journals. He holds many copyrights and patents and has organized many national and international conferences. He is associated with several universities as an external guide for research scholars and has conducted many workshops at institutions.

Shalini Aggarwal, PhD

Associate Professor, Chandigarh University, Ajitgarh, India

Shalini Aggarwal, PhD, is working at Chandigarh University, Ajitgarh, India, as an Associate Professor. She has a working experience of more than 18 years. She has presented various papers at conferences, including those at the Indian Institutes of Management at Lucknow, Bangalore, Shillong, and Kashipur, and has published 24 papers and four case studies, several of which have been indexed in Scopus and and Australian Business Deans Council (ABDC) listed journals. She was selected as a reviewer of the Emerald journal *Journal of Financial Economic Policy* and has reviewed around 30 papers. Recently Dr. Aggarwal has been appointed as a member of the Technical Program Committee (TPC) of AF 2019 for the 9th Annual International

Conference on Accounting and Finance (AF 2019) from 15–16 July, Singapore. She was also appointed as the guest editor of a special session entitled "Green IT and Sustainability" of the International Conference on Innovative Computing and Communication (ICICC–2019), organized by Technical University of Ostrava, Czech Republic. Many researchers are doing research under her guidance. She also completed one minor research project entitled "Impact of Self Help Groups on Social, Economic and Political Empowerment of Women: A Case of Ghaziabad District," financed by the Institute of Management Studies, Ghaziabad, India. Dr. Aggarwal's degrees include an MBA, PhD, and MSc in Computer Science.

Pooja Tiwari, PhD

Faculty Member, ABES Engineering College, Ghaziabad, India

Pooja Tiwari, PhD, is a faculty member at ABES Engineering College, Ghaziabad, India. She has over 12 years of experience in teaching and research and as well as industrial experience in production. She has many national and international publications to her credit, including papers in Scopus-indexed journals. She takes an active part in seminars, conferences, and organizing workshops. Her area of interests is human resource management, cross-cultural management, etc. She is an active member of various professional bodies and also a member of editorial boards of many international journals.

Prasenjit Chatterjee, PhD

Associate Professor, Mechanical Engineering Department,
MCKV Institute of Engineering, India

Prasenjit Chatterjee, PhD, is currently working as an Associate Professor of Mechanical Engineering Department at MCKV Institute of Engineering, India. He has over 80 research papers in various international journals and peer-reviewed conferences. He has received numerous awards, including a Best Track Paper Award, Outstanding Reviewer Award, Best Paper Award, Bright Researcher Award, Outstanding Researcher Award, and University Gold Medal. He is the guest editor of several special issues in several

Scopus-indexed journals and Emerging Sources Citation Index (Clarivate Analytics) journals. He is currently editing and authoring books on decision-making approaches and sustainability. He is the Lead Series Editor of International Perspectives on Decision Analysis and Operations Research book series (Emerald Group Publishing) AAP Research Notes on Optimization and Decision-Making Theories (Apple Academic Press); and Frontiers of Mechanical and Industrial Engineering (Apple Academic Press). Dr. Chatterjee is one of the developers of a new multiple-criteria decision-making method called "Measurement of Alternatives and Ranking according to COmpromise Solution" (MARCOS)..

CONTENTS

CONTRIBUTORS

Shashank Agarwal
Finance Officer (Head of Finance and Accounts), Kasiga School, Dehradun, Uttarakhand, India, E-mails: shashank0711@icloud.com; financeofficer@kasigaschool.com

Priyanka Bhardwaj
Chandigarh University, Punjab, India, E-mail: b.piya16@gmail.com

Sharmistha Dey
Maulana Abul Kalam Azad University of Technology (WBUT), Kolkata, West Bengal, India, E-mail: papri.dey@gmail.com

Ankita Dhamija
Lingaya's Lalita Devi Institute of Management and Sciences, New Delhi, India

Richa Goel
Amity University, Noida Campus, Uttar Pradesh, India

Divyansh Gupta
Birla Institute of Technology and Science, Pilani, Rajasthan, India

Sachin Gupta
Assistant Professor, Department of Business Administration, Mohanlal Sukhadia University, Udaipur, Rajasthan, India, E-mail: sachinguptabusadm@gmail.com

Om Prakash Gusai
Assistant Professor, Department of Commerce, Motilal Nehru College, University of Delhi, New Delhi–110021, India, E-mail: opgusai@mln.du.ac.in

Priyanka Jingar
Research Scholar-Doctoral Studies, Department of Business Administration, Mohanlal Sukhadia University, Udaipur, Rajasthan, India

Srishty Mehra
University School of Financial Studies, Guru Nanak Dev University, Punjab, India, E-mail: srishty012@gmail.com

Gaurav Nagpal
Birla Institute of Technology and Science, Pilani, Rajasthan, India, E-mail: gaurav.nagpal@pilani.bits-pilani.ac.in

Shipra Pathak
Chandigarh University, Punjab, India

Annu Purthi
Chandigarh University, Punjab, India, E-mail: anupruthi0911@gmail.com

Ankur Rani
Research Scholar, University of Delhi, New Delhi–110021, India

Seema Sahai
Amity University, Noida Campus, Uttar Pradesh, India, E-mail: ssahai@amity.edu

Ravi Sanwal
S. V. Government PG College, Lohaghat – 262524, Uttarakhand, India, E-mail: ravisanwal6@gmail.com

Deepshikha Seth
Amity University, Uttar Pradesh, India

Deepa Sharma
Faculty and Research Scholar of Maharishi Markandeshawar (Deemed to be) University, Mullana, Haryana, India, E-mail: deepasharma36099@gmail.com

Gurmeet Singh
Head of Graduate School of Business, University of South Pacific Statham Campus, Fiji

Shalini Srivastav
Amity University, Uttar Pradesh, India, E-mail: ssrivastav@gn.amity.edu

ABBREVIATIONS

ACFE	Association of Certified Fraud Examiners
AGI	artificial general intelligence
AI	artificial intelligence
ANN	artificial neural networks
ANN	automated negotiation model
ASI	artificial superintelligence
AVE	average variance extracted
B2B	business-to-business
B2C	business-to-consumer
BARO	best analysis and rapid outcome
BCG	Boston Consulting Group
C2B e-commerce	consumer-to-business e-commerce
C2B	consumer-to-business
C2C e-commerce	consumer-to-consumer e-commerce
C2C	consumer-to-consumer
CBR	case-based reasoning
CRM	customer relationship management
CSR	corporate social responsibility
CYBG	Clydesdale and Yorkshire banking Gathering
DHCP	dynamic host configuration protocol
EA	evolutionary algorithms
EDI	electronic data interchange
EDP	eco-domestic product
EIS	Earley Information Science
F-commerce	Facebook commerce
GBM	gradient boosting method
GDP	gross domestic product
GLM	generalized linear model
GNP	gross national product
GPU	graphic processing unit
HKFI	Hong Kong Federation of Insurers
IRDAI	Insurance Regulatory and Development Authority of India
KPIs	key performance indicators

MAUT	multi attributive theory
MI	machine intelligence
MIDAS	motor insurance DLT-based authentication system
ML	machine learning
NLP	natural language processing
NN	neural networks
PPP	polluter pays principle
PwC	PricewaterhouseCoopers
SCM	supply chain management
SCM	supply chain marketing
SEEA	system of environmental and economic accounting
SEM	structural equation modeling
SNA	system of national accounts
SNSs	social networking sites
SPT	social presence theory
VCM	value chain management
WIPO	World Intellectual Property Organization
WM	warehouse management

ACKNOWLEDGMENTS

We wish to express our sincere appreciation to those who have contributed to this book and supported us in one way or the other during this amazing journey.

First and foremost, we would like to thank with utmost devotion and humility *"The Almighty Lord"* for clearing all the obstacles and making it rather an opportunity than a problem for this valuable work.

We would like to acknowledge the immense role of our families in enabling us to complete this project. We consider ourselves the luckiest in the world to have such a supportive families standing behind us with their love and support.

Many people played an important role in the development of this edition of the book, and we are deeply grateful to all of them.

—Editors

PREFACE

In the current competitive and technology-driven world, AI is an emerging technology, and it is considered a big deal. AI is the study of the design of intelligent computational agents. This book is written as the editors of this book were quite fascinated by the emergence of artificial intelligence (AI) as emerging science and technology. It will be quite useful as we have tried to provide depth by presenting different research papers and in-depth results. We have to cover not only the issue related to the domain of finance. Other general applications of AI in other domains are also taken into consideration.

Our main focus of this book is to explore the ideas that have emerged in the past few decades of AI, and also the work done by the various researchers is presented. We have tried to keep this book quite precise and tried to maintain less formality in the presentation.

This book will be quite useful for people in the field of management and also for the people in the domain of technology. Due to its coverage, it is quite useful for students and professionals in the area of AI.

In addition, the endeavor of this book is to create sky-scraping quality, highly trained, and highly skilled graduates. This book is one such initiative of a collection of cutting-edge, high-quality chapters backed by thorough and systematic research.

We specially thank the diverse and international review board members for their special review. Feedback and suggestions are sincerely invited from our readers to help further improve the quality.

We hope that this book enriches its readers with contemporary areas of international business and gives an understanding of the best practices from around the globe. We wish readers an enjoyable reading experience!

—Editors

INTRODUCTION

AI is used in every walk of life. It has transformed the way people used to assimilate information, analyze data, and make decisions. Various financial companies, media companies, and insurance companies are finding a way to use AI to simplify operations. The big giants like Apple, Google, and Amazon invest huge amount of money to incorporate AI into their business to gain maximum benefit.

The reach of AI is extensive. It can be used to detect fraud, write news stories, handle customer interaction through chatbots, and explore space. It also can be used in education, transportation, and much more. For instance, the use of AI in the stock exchange brings wonder. All the buy and sell transactions are matched within no time using a computer and without human intervention. Robots analyze the market data to predict a change in the stock market and manage finances as compared to human counterparts. Similarly, storing information on the cloud has become a reality that saves space and time.

To maintain a competitive edge with respect to customer services, Amazon uses machine-learning technology. It helps customers to find products for themselves. By using collaborative filtering and next-in-sequence models, Amazon can predict what the customer might want next. Further, Amazon uses AI in the logistic business as well. AI is used to schedule the timings of product delivery and for accuracy and efficiency. Amazon is trying to make possible the use of delivering goods to customers at their doorsteps with the help of drones.

In India, almost $150 million have been invested by private players alone in the area of AI since 2016. As every year transactions through the internet and online banking are increasing, there is also an increase in theft in the banking and financial industry. Therefore, it becomes important to have thorough knowledge of AI and its applications. AI can bring financial cybersecurity to the next level by using deep learning technologies. The use of blockchains can prevent fraud. Once the document written in the blockchain is locked, it is stored forever. It cannot be tampered with or manipulated. The data is saved.

e-commerce business is growing at a fast pace. It has been made possible by embedding AI into business. Amazon and Apple have developed Alexa and

Siri, respectively, that are conversant with humans and try to find a solution to their queries. They allow humans to communicate with a search engine using their voice commands. Similarly, chatbots provide 24/7 assistance. They help in understanding the needs of customers and help businesses to develop products accordingly. Consumer interaction through chatbots leads to a satisfied customer, thus increasing productivity and earnings for the company.

The use of AI in teaching is very impactful. Notes of professors conversing in one language can be translated globally into different languages by embedding AI. PowerPoint has the free plug-in "Presentation Translator" that generates subtitles in all the languages for what the teacher is teaching. Even visually challenged people and hearing-impaired people can use it. The virtual classroom is made possible through the Zoom software. This has an added benefit of allowing a large number of students and teachers to connect at the same time.

Sustainable development is the need of the hour. The use of AI in business promotes sustainability. Development of green accounting, green HRM, and green marketing have been made possible using AI in business. Routine tasks of managing finance can also be done by using the technology of AI, such as assessing credit quality or automating client interactions. These are some of many financial services that have resorted to the technology of AI in storing and analyzing their data.

The scientist is waiting for the day when AI can be used to its full capacity. It will be a time when machines will surpass human intelligence. It is believed that it will reinvent the world as it is today and the machines can perform all the menial tasks. Humans can then concentrate on working at higher levels of work to accomplish a lot more.

CHAPTER 1

IMPACT OF AI ON E-COMMERCE

DEEPA SHARMA

Faculty and Research Scholar of Maharishi Markandeshawar (Deemed to be University) University, Mullana, Haryana, India, E-mail: deepasharma36099@gmail.com

ABSTRACT

Computer-based (AI) intelligence will double the pace of advancement within India's business associations. Artificial intelligence (AI) is helpful to enhance economic growth and every person's standard of living. In Indian business, AI does an indispensable job in the future occupation. Organizations and people groups are eager to put resources into human capital, or e-business plays a vital role in providing quality goods and services to consumers in minimum time. Furthermore, expanding on AI prepared workforce doesn't interpret into an intense requirement for mechanical aptitudes. E-commerce has a positive impact on our life. It's like a revolution in how we learn and establish our business and run online, etc., and how we work. In competitive period organizations wants to earn more and more profit to produced qualitative goods for consumers. On the other hand, consumers fulfill their requirements, needs at home without wasting time. AI fulfills both party's objectives. AI solves out the multiple problems of business people and consumers. Computer-based intelligence gainful for organizations, managers, purchasers, or economic development too. AI undoubtedly has a positive impact on human being lives. We have been offering quality types of assistance to worldwide customers for over the years now. Our never-ending experience of having worked with internet business customers from various spaces gives us an upper hand over different players right now. We have a group of exceptionally talented information section officials who are prepared to deal with any sort of online business information passage administration demands. Our rehearsed approach, brisk circle time, tender loving care, and energetic help guarantees that all your business needs will be specifically met and your internet business organization will be route in front of the opposition.

1.1 INTRODUCTION

In today's era, every person influences by artificial intelligence (AI). AI is used in every field, e.g., education, business applications including, manufacturers, operation process of the product, marketing, to analyze consumer behavior, data recording about selling the product, to comparison with competitors, natural language processing (NLP), and across industries. AI plays a vital role in the competitive world. Through the AI to solve the bigger problems in nano-second. Every business person captures the market by producing a quality product and identifying the consumer needs, wants, and desires a selling the product. All activities are completed to consumer needs. Because customer satisfaction is the basic priority of every producer. Consumer behavior is observed by every businessman when he wants to first time familiar with his new product in the market. The consumers about their product adopt the many methods, techniques, and technology for it. E-commerce plays a vital role. Because with the help of e-commerce, marketers easily delivered the product in the hands of consumers and consumers save the time easier to get the product through an online payment. AI helps the e-commerce business to easily capture the market and earn a profit and timely fulfill the consumer's wishes, needs, wants, and desires [1–23].

AI and e-commerce collectively used by the businessman to increasing sales because AI includes human beings' intelligence to be used the machines and computer systems. Now our new generation saves time gives the most importance to electronic transactions, and AI helps to provide more and more facilities to the consumers. AI helps to adopt an effective and economical manner and procedures for manufacturing quality products.[1]

1.2 POSITIVE IMPACT OF AI ON E-COMMERCE GROWTH

AI and e-commerce have been used by businessmen and customers satisfaction to be creating a benefit for them. Recently AI and e-commerce both are uncompleted to each other. E-commerce has purely based on

[1] The impact of AI on e-commerce development.

consumers' timely changes in trends, customs, designs, and styles changed but with the help of AI, creative new ideas to identify consumers buying behavior and, according to consumer's demands, manufactured the products. Every business wants to earn them more and more profit; Goodwill and profit are based on sales of the goods and services through the promotional tools and adopted effective advertising strategies for selling the product. The marketing strategy part of e-commerce. AI and e-commerce combine used by a company to reach its target market through various ways like through media, advertising, and effective promotional tackle, etc.

1.2.1 FOUNDATIONS OF AI

AI is a great discovery in today's world. AI makes every person's life too easy and comfortable, especially those who link with technology. AI is blessed for every human being who is touched by AI and e-commerce. AI is the base of development of e-commerce development. AI is based on some foundations (Figure 1.1).

FIGURE 1.1 Representation of AI-based foundations.
Source: By author.

1.2.2 AI IMPACT IN E-COMMERCE

AI influences e-commerce because customers are always the king of the market. Therefore, according to the customer's convenience, more preferences are given for online shopping. AI also blessed for an illiterate person who does not know how to write but through the voice search option place the order and fulfill their demands and desires and new items according to their needs, wants, and desires without using the laptop, Android phone more convent to consumers to do shopping without wasting time and do the shopping through e-commerce in economy manner.

E-commerce also helps the marketer to analyze consumer-buying behavior because of which type of consumer needs, demand about the particular product a manufacturer understood and produced. In addition, satisfy the consumers and marketers easily capture the market and meet the competitions. Sometimes consumers do not want to buy particular products, but with the help of digital marketing methods and advertising, the media helps the company to influence the consumers for goods to buy. Creating the apps, it's a part of the marketing strategy that permits clients and prospects to obtain a particular product from exacting situate. AI is an economical manner to attract customers to buy a particular item for consumption. Consumers to buy a particular product according to their needs, wants, and desires. Not only buy the product easily also to compare the one particular product price to other company's products. Now the company gets any feedback from consumers to make efforts to improve the quality of products and through the advertising convey the message about the product, and which types of testing go through by the products (Figure 1.2).

- **Recommendations Engines:** AI fulfills every person's desires, needs of every human being. Producers produce the products, as a result of manufacturing the goods employees need and finished goods to deliver the consumers' door human being needs. Therefore, as a result, employed a person like Amazon, snap deal, too many other companies employed too many people to sell the products and attract customers day-by-day new schemes launched and persuade the consumer to buy the particular things of the company. Sometimes consumers ready to buy the goods, but he has no enough time to purchase the goods; AI help those consumers. They are fully satisfied with their requirements and desires.
- **Virtual Assistance:** AI is also helpful to make a list of those students who buy the products and place the order for delivering the goods to their homes. The consumer buying behavior about to purchase the specified goods and services easily analyzed through AI.
- **Particular Product:** These are help full to predict the prospects who easily convert customers. All reputed companies adopt that method and techniques to analyze consumer behavior.
- **Based Chatbots:** AI in e-commerce not to be complete without a discussion about chat boats. AI plays an important role in e-commerce because consumers easily visit business websites or links to gather

all information about products and company history. Chatbots also a medium to provide comfort to consumers along with the quality product. So, as a result, if consumers like the company product easily, companies goodwill make.

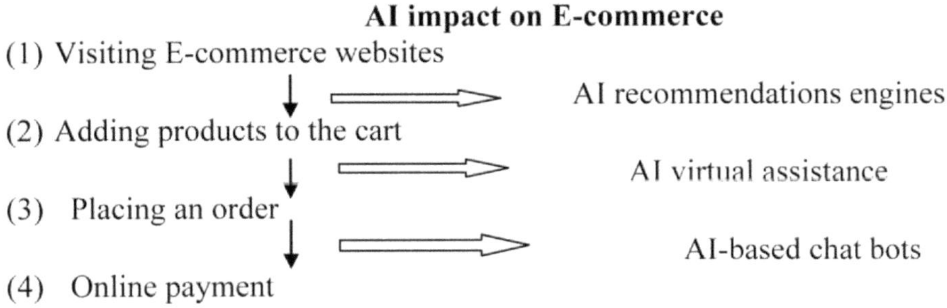

FIGURE 1.2 AI impact on e-commerce.

Source: By author.

1.3 SUCCESSFUL UTILIZATION OF ARTIFICIAL INTELLIGENCE (AI) IN THE PRESENT ERA

AI creates a new path for the development of a business hub. In the modern era, competition is too high, and every businessman wants more and more to capture the market to formulate proceeds. Every company that's successfully integrated AI technology into multiple aspects of its business, a consumer goods corporation. With the help of AI forecasts of marketing analytics and AI also helps to forecast the purchaser activities forecasting about invention preferences.[2] AI is used in fraud detection, personalized advertising, and marketing messaging and customer services, etc. (Figure 1.3).

e-commerce is very valuable for the consumers and company because e-commerce and AI both are equally used by both parties easily achieve the goals that mean consumers with the help of AI save their time and buy the goods easily at home, and companies provide the unique features of the product to consumers enhance their goodwill and easily meet to the competition. In addition, capture the whole market. Competitions facing many companies who do online business like Mantra and Snapdeal use the double algorithms and appoint the workers on the job, on the work to check their

[2] How big data and AI are influencing the e-commerce industry.

efficiency who doesn't achieve their target easily. AI maintains all records related to them. AI also helps to 9 to 5 workers jobs.

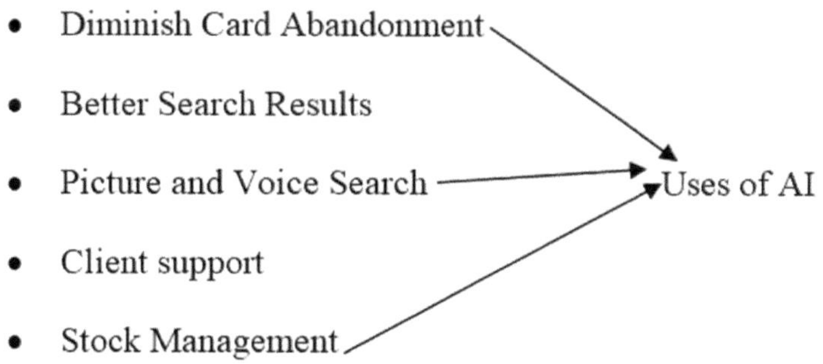

FIGURE 1.3 Representation of some uses of AI in the business hub.
Source: By author.

1.4 AI IS A FOUNDATION OF E-COMMERCE

AI is very beneficial for all companies because consumers always the king of the market, so fulfill the consumer's needs, demands companies want to know what are needs and desires of a particular consumer and why a consumer buys a particular company product. To expand the business to persuade the consumers to buy the manufactured goods. AI always works to support producers, businessmen, and consumers smoothly to do the production process and capture the market.

1. **Augment Search:** Search is a significant piece of web-based business and AI. Through the assistance of search, devices convince the intentional interest group to purchase a specific item. It's useful to make new thoughts and possibilities for clients because the organization as per the purchaser's needs and wants to create quality commodities and ventures and catch the market lacking any problem.
2. **Automate Supply Chains:** AI helps to supply should be enhancing according to the consumers' necessity for manufacturing the goods. As a result, it easy to maximize the profit and minimize the cost.
3. **Plan for Tomorrow:** Man-made intelligence accommodating in estimating the future, to deliver quality merchandise to dazzle the clients to purchase a specific organization item. Their simulated intelligence assists with anticipating the future, as shoppers want to fabricate

quality items and supply the products to satisfy the needs. Convince clients to purchase a specific organization item, and AI assists in improving the potential and retailers hand-to-hand to fulfill needs.

1.5 AREA OF ARTIFICIAL INTELLIGENCE (AI) APPROPRIATELY USED

AI assumes an exceptionally fundamental job in the present time. Man-made intelligence is nearly utilized by each field like the Education office, Computer science, Medical field, job search, Human Resource planning other fields also in touch of AI. Some of the fields are discussed in subsections.

1.5.1 COMPUTER SCIENCE

An AI developer also helps to launched and developed effective methods and techniques to solve the particular problems related to computer science. AI had done too many best ways to solve difficult problems.

1.5.2 EDUCATION

The role of AI in the study is commendable because AI fully supports the education sector and to be used too much and effectively to guide the students through modern technology. Some benefits of AI to make easy tasks for the students. AI helps students to query all the typical questions related to the study. Today is technology time easy the student's life. Through AI, students learn more and more things and easily doing the study very covalently. There are likewise frequent basic and troublesome issues by AI a few times. AI would play a very important role in our education sector if someone understood the AI program. AI helps everyone with problems if they have. Through the AI-enhanced, the knowledge of students and give them better and better options to solve the typical questions. AI also helps the students to make the slides; as a result, to help the slides, students easily understood the relevant point, and they furthermore utilized the slides to be given the presentation about their topic. AI also helps to 3D alignment, graphic representation to make the slides effective and memorized also. AI should be helpful to the teacher; it reduced the paperwork cost, through the slides clear the student's doubt, and students also show interest in the topic.

Teachers also used projectors, Laptops, notebook, and graphics and images to be used to clarify the student's concepts and clear the doubts. For the teachers, AI blessed the students and teachers also. Teachers easily clarify the student's concepts and doubts. Understudies and instructors through AI move the information from one spot to another place. AI can provide our many features to a classroom and easily take the queries about our doubtful questions. However, many teachers do not like AI because they never come to their comfort zone. They don't embrace new innovation since they feel frenzy to receive new things and implement their professional life.

1.5.3 UNDERSTOOD MARKET ENVIRONMENT

It's not easy for a businessman to capture the market and persuade the clients to be invested their money in their projects. Many methods and ways are adopted by many companies to convince the clients to help to make their investment decisions—all financial decisions to make easy through the AI. Advertisers in day-by-day schedule make various counts and store and record the information of each client, it's not an easy task to manage the records of clients but through the AI very helpful for financing and marketer person also. Its great options or alternatives of functions include many natural languages; if somebody is from another state, his language is differing from other people, but through the AI, he easily communicates with the persons. Computer-based intelligence depends on AI frameworks that effectively feed the numerical numbers and get the result.

1.5.4 FINANCE

Many goods and products are produced by the manufacturer because the producer wants to profit. When they produced the product to introduce their product, they used too many strategies to capture the market. First of all, manufacturers study the market, analyze consumer behavior and other things. According to consumer needs, desires, and wants, they produce goods and services. Man-made intelligence is used as an apparatus to finish all the above assignments. Everybody who set up his business to make a benefit is the principal objective. The customer is the ruler of the market. To present the item and convince the client to purchase a specific item promoter a lot of exertion put through the man-made mental power, therefore, helps shoppers when deals the items since customers going through their cash to buy the commodities and enterprise. Man-made intelligence can, without much of a stretch, distinguish all issues identified with buyer conduct, money uniformity, and examination of why buyers spending their cash on a specific item and why customers take definite purchase items.

1.5.5 GOVERNMENT

While the government will no doubt benefit from AI, it will likely also play a job in humanizing the knowledge by sustaining efforts to manage the potential for bias within the algorithms AI systems use in all government sector related to employees data and AI-based functions correctly and provide the fast and accurate data about populations as result government work in favors of peoples.

1.5.6 HEAVY INDUSTRY

Robots are the latest example of AI. Robots play a very important role in industries as home. Now Robots take the positions of human beings; in other words, we can say that robots take place the human jobs. Robots are a significant disclosure of AI. Robots, not arguments like a man and never tried, and full of emotions like a human being. Robots are a danger for human beings. No doubt, robots are also effective or efficient—electronic man (robot) to make mistakes when humans or programmers filled the wrong data. Robot's good work is done without tiredness and baseness.

1.5.7 HOSPITALS AND MEDICINE

In emergency clinics, AI assumes a significant job to make body X-beams of patients and collect the money to receive from patients automatically and easily found the bone age through the computer software. Through the AI easily perform the tasks and easily maintains the records and also maintains records of the medicine stock.

1.5.8 INTERPRETATION

AI also helps to interpret the information in digital images or graphic images and highlight important things. AI helps the doctors to listen to the heartbeat sound of patients since AI completely bolsters the human body and medical areas. Easily make appointments and consultancy schedules.

1.5.9 HUMAN RESOURCES AND RECRUITING

Human resource managers easily make the planning, E-HRM, recruiting the applicants, analyzing the shortage and surplus of the human being as employees in the organizations. Also, getting contrasts between trademarks is unmistakably progressively complex in view of the fact that it is a lot harder to discover sets of comparable trademarks, and in addition, there is no predictable meaning of in the same way considered being comparable just in the event that they are deceptively proportional. To add another component of intricacy to this issue, we should consider what happens when we have two trademarks with fundamentally the same as figures yet with various content. In the event that we take a gander at the content reproduction of these trademarks, we will get awful outcomes. We can't overlook the content, as it may be scripted in the trade name picture itself or aside some other figure which takes after another trademark. These inquiries drove us to the idea that we need to part the comparability question to various angles, each with its focus and its solution.

1.5.10 JOB SEARCH

The employment bazaars have seen a talented candidate to their vacant seat in the companies. We also implement AI in the area of job search. AI is simply beneficial for both parties, like job seekers and recruiters. The neural network was given many images showing diseased and healthy plants. The team is planning to add more human talent in the organization and offer to join the capable company candidates and to be used their skills, knowledge, skills, and capabilities; as a result, to enhance the goodwill of the company. The employee's skills must be learned, like spotting signs of a disease in a video or medical scan in the medical sector. AI maintains the record about their work, absenteeism, and contribution in task performing should help to company because of promotion or performance appraisal of an employee based on the above things. Therefore, AI assists in making a simple assignment to advance workers in the organization. Man-made intelligence likewise causes coordinating individuals as indicated by their intrigued region and ability premise to move on their relevant posts and positions. Machine intelligence

(MI) calculates the wages and salaries of individual employees to perform a particular job and bold the main information about the resume. HR manager to meeting position on their farm duties in privileged fondness.

1.5.11 MARKETING

Synthetic brainpower has been making large news about all sectors and fields. Its live example is digital marketing. AI is a trending topic in today's era. Many candidates are making their career in digital marketing have many options created by AI. AI is a great discovery in today's world. AI became a very easier life for consumers, businessmen, and everybody who links with technology. Through the use of AI and e-commerce to do online shopping and save time.

1.5.12 MEDIA AND E-COMMERCE

A quantity of AI applications is put into practice on medium comfortably resembling such as movies, TV programs, advertisement videos, audios, etc. AI also plays a vital role in it because when consumers do online shopping, voice search allows customers to search for new items according to their needs, wants, and desires without using the laptop, digital media. We live in a disturbing world, where we have a lack of time to go to the market and purchase appropriate retail for living. Organizations that are coordinated, regulating, and exceptional will do well, and the rest will die. Computer-based intelligence causes organizations to have a place seen for their job in following execution and proficiency are presently accused of through and through the system, and as impetuses for change development. They should keep up their organizations and budgetary remaining as well as fill in as change specialists also. They need insight to consolidate innovation adequately and the intelligence to reframe innovation from being ideal to have capital consumption to an operational cost. Digitization is currently a practical objective for the account work due to a scope of innovative advances.

1.5.13 MILITARY

Accessibility of recommendation engines has through its uncomplicated for e-commerce arrangements to customize the shopping experience for clients.

In light of the purchaser's past perusing examples and shopping subtleties, the internet business site can show custom-made item proposals that are related to the purchaser's advantages. Another way personalization helps with dynamic estimating. Since the typical request, an incentive for a B2B client is higher contrasted with stores can execute lively cost to improve the deal. One all the more methods to do this is in a topographical setting. Stores can show items and offer that are increasingly suitable to the spot of the purchaser. While we are discussing personalization, dealers can likewise suggest guarantees, extra trimmings, etc., using AI.

1.5.14 MUSIC

Simply a month ago, the World Intellectual Property Organization (WIPO) reported the dispatch of another picture search innovation controlled by AI. It will be actualized through trademark workplaces around the world, covering very nearly 38 million imprints as a feature of WIPO's Global Brand Database. Dissimilar to past cycles of AI-controlled picture search, which for the most part, depending on the investigation of shape or shading to decide likeness, the new instrument can distinguish blends of ideas to do likewise.

1.5.15 NEWS, PUBLISHING, AND WRITING

Correspondence, in addition to a joint effort between IP workplaces, is probably going to be pivotal for AI-based answers for being actualized viably. Director of the Global Databases Division at WIPO. He says current joint efforts with national workplaces furnish WIPO with an assortment of information that is aggregated, arranged, and cleaned for building AI apparatuses. The Madrid System has been utilizing machine interpretation to accelerate crafted by interpreters for merchandise and venture writings, decreasing interpretation-redistributing costs. As the AI innovation is open source, WIPO can expand on and redo it to meet the association's requirements.

1.5.16 ONLINE AND TELEPHONE CUSTOMER SERVICE

In time AI will improve the nature of choices, both by upgrading the limit of the office and the nature of indexed lists by an official; the innovation permits to disparity picture trademarks and staggering precision, discovering every single significant picture inside its database that contains a similitude, clarifies, trademarks, and geographical indications. It additionally makes the procedure speedy and hard. Having the option to lead this stage consequently would be valuable to save time and speed up the registration process.

1.5.17 POWER VALUE

Force is truly significant in this serious condition; while there's unquestionably far to go before AI can upset crafted by IP workplaces, thc spccd of hcadway proposes it is imprudent to foresee that it won't. Meanwhile, the innovation is, as of now, upgrading IP workplaces' work, and maybe more fundamentally, it is demonstrating progressively helpful for clients of the frameworks. That pattern is probably going to proceed as more workplaces receive the innovation, and explicitly, as activities. AI arrangement is predicated to an enormous degree on the amount and nature of accessible information. The degree to which IP workplaces organize their endeavors here is probably going to have far-reaching repercussions for the fate of the business. On any occasion, the times of trademark inspectors physically trawling through a huge number of utilizations are, on the off chance that not finished, at that point in transit out.

1.5.18 SENSOR

Furthermore, an AI-prepared workforce doesn't really convert into an intense requirement for innovative abilities. The best three most required abilities among business pioneers in India effectively catch the market. Some of the time, purchasers would prefer not to purchase specific items. However, through the noteworthy limited time, apparatuses ought to be utilized to convince the customers.

1.5.19 TELECOMMUNICATIONS

Looked for a veterinarian in your general vicinity, and now your social feed has been taken over by advertisements for pet items? That is the intensity of AI. Through AI, retailers can find dynamic examples in client habits and provide bespoke shopping experiences. With the power of machine learning

(ML), retailers can now even predict customer choices even before the customer knows about their requirements. Thinking about buying pet food and suddenly pet food advertisements are all around you? Well, it's not sorcery.

1.5.20 EFFECTIVE TOYS

An e-trade store's plan and nature of item pictures assume a basic job in driving deals. Creators have embraced responsive structures to adjust sites to different screen goals throughout the years. In any case, with the utilization of ML, fashioners are presently endeavoring to customize the site to singular clients. Visualize a store organized and showed to you! With AI, style e-posteriors are additionally now moving endlessly from customarily costly photograph shoots and utilizing AI to take care of the clothing photography issue. Picture altering administrations, for example, utilize AI to make completed pictures inside minutes with practically no human mediation by any draw out of the imagination.

1.5.21 TRANSPORTATION

With the assistance of critical edge evolutionary algorithms (EA), retailers can anticipate content promoting results through testing gatherings of potential arrangements. With dynamic learning, more gatherings of arrangements are consequently evolved and tried, in light of the achievement of past gatherings. Brand informing, site plan, and design, the proposal motor are a portion of the key regions can effectively reshape.

With the headway in AI and better equipment, the online business experience is going to change radically for both purchaser and vendor. The inquiry is on the readiness of E-retailers to take advantage of this lucky break.

1.6 AI IS SPREAD OF E-COMMERCE

AI concept is a very popular concept and tool of e-commerce since to expand item deals and catch the market, AI extremely considerate. AI is a fact of today's world. In day-to-day life, daily new products launched by the companies and people so busy in their own fast life. They have no time to go market and buy things. AI and e-commerce, both concepts used together, create a history. Every company launches the daily biggest brands and influences the customers to procure exacting manufactured goods. To make the successful implementation of brands AI and e-commerce, both tools are used together in today's era.

1.6.1 CUSTOMER EXPERIENCE

Customers have a great experience with AI and e-commerce. The client is the lord of the market. To temporarily deprive of sight or pull in clients to create products and ventures. Customer loyalty is most important because dependable customers spend all their money and savings to buy thc goods and services to satisfy their needs, wants, and desires. Easily make the new customers by the company.

1.6.2 SEARCHER INTENT IS KEY

A little that AI systems are receiving a large amount superior at treatment is accepted language and how people usually speak as opposed to how they state things when they utilize web indexes. This is so expensive because the searcher goal is massively important. What are people looking for?

1.6.3 SHOPPER INTENT IS KEY

A customer looks for buyer conduct about a specific item. Why utilizing its division driving proposal framework, and each wholesaler should utilize them. A framework on condition that an AI drove suggestion attracts from client movement to choose the things designed to entice the customer, prompting more noteworthy request esteems, and then some did they want to buy it? What do they want? A consumer buys goods for need, sometimes for status symbols like a bike, motorbike, or barrel kit? The quicker you know, the better for you can supply it. The perceptive searcher surroundings and general trends, AI search, can make results more significant.

1.6.4 SMART REVENUE

A Superstar through the advertising influence and persuade the customers or their fans to buy a particular company product. A company that hires a superstar when a famous person reaches the depart position should be drawn absent all the stops to maximize their request esteem. This is done through a combination of up-selling and cross-selling and the former offering higher-end versions of chosen products, and the latter contributing like-minded products. Amazon introduced smart recommendations to the typical community returns.

1.6.5 CHATBOTS ARE READY FOR THE HIGHLIGHT

A lot of sellers have previously added chat genuine individuals physically field inquiries throughout the day is a decent method to bots to their provisions since they are greatly versatile for purchaser maintenance. Have frame-up operating costs, plus citizens can't scale and but chatbots can. Visit bots can likewise be second-hand to gigantic outcomes in territories outside of buyer support to enhance the management boundary of your store, for instance. Flashily lead the way in this look upon when it acquired the Kit chatbot, modified it, and complete it free of charge for anyone; a trader can alter their item costs and run promotions, all from the appearance book dispatch.

1.6.6 FORECASTING THE DEMAND AND SUPPLY

AI tackle is outstanding at forecasting the demand of consumers and measures earlier results to estimate prospect results, and anticipating is a crucial piece of stock control. In the event that you can. Settle on what point of command await and can optimize your supply levels.

1.6.7 OPERATIONAL EFFICIENCY ALWAYS GROW

A lot of the improvement AI determination transport motivation obtains rest following the scenes in the day-to-day pound because it is out of the ordinary at achieving prepared good organization. The efficiency comes behind judgment, and central processing unit systems are essentially reason equipment. The world is using information by no means, and the conditions figures discipline and statistics analysis is worn approximately interchangeably today. Abundant citizens believe that an information scientist is just an imagined surname for a records analyst (Table 1.1).

TABLE 1.1 Huge Use of AI in India

The Weather Channel	(forecast)
Amazon	(packages)
Google	(searches)
Giphy	(service up)
Netflix	(users use)
Snapchat	(snaps)
YouTube	(videos)
Skype	(calls)
Instagram	(users post photos)
Americans	(of internet data)
Tinder	(users see match)
Uber	(users take rides)

Source: By author.

1.7 E-COMMERCE TREND AND ARTIFICIAL INTELLIGENCE (AI)

The conventional e-commerce search is principally motivated by content, yet with the approach of AI and ML, purchasers will no longer need to depend on a content-based pursuit. Facial appearance like text-to-speech can speed up the buying process. You can utilize computerized colleagues to checked search—the ease of use of the product and place an order. An additional oversize step, in requisites of technology, in this area is image acknowledgment. Want to buy a product but don't know the brand name or its industrial condition? Just click a photo and upload it to the application. The AI calculation will carry out the responsibility for you by coming back with the most applicable products.

1.7.1 IMPROVING PRODUCT AND PRICING

Another way personalization helps with dynamic estimating. Since the usual order value for a B2B customer is higher compared to a B2C customer, stores can execute energetic prices to make the sale better. Stores can show products and offers that are more appropriate to the spot of the buyer. While we are discussing personalization, dealers can equally prescribe guarantees, extra trimmings, etc., using AI.

1.7.2 PRICING OPTIMIZATION

In the period of statistics, every companionship has calmed approximately every geographic, demographic, and behavioral measurement and metric. They also have exhaustive obtain data from their company. Despite this, the company is unmoving, not competent to make the most of these facts—your buyer personas outward appearance the foundation for the deals and showcasing system. Simulated intelligence, you can crawl through this information focuses on improving your purchaser personas and portioning your clients in a superior manner.

1.7.3 IMAGE TAGGING/RECOGNITION

Simulated intelligence makes another way for the advancement of a business center point. Each foundation in the present period wins an ever-increasing number of benefits in contrast with its Competitors. Each organization that is effectively incorporated AI innovation into different parts of its business, a purchaser merchandise partnership. With the assistance of AI gauges of promoting examination and AI as well assists with anticipating purchaser conduct and forecast about item suggestions.

1.7.4 WEBSITE PERSONALIZATION

AI will dramatically increase the pace of development inside business associations in India and agree on business pioneers in India. The discoveries, which are a piece of an examination led by Microsoft. Now and again, customers would prefer not to purchase specific items; however, those are impacts for looking for merchandise with the utilization of pictures. Man-made intelligence to make the systems to make the social changes and skilling are requirements for AI to work. Computer-based intelligence is a centerpiece of their system and builds up a learning spryness culture. Investment in this transformative innovation to be persistent for drawn-out progress.

1.7.5 CHATBOTS WITH VOICE ASSISTANTS

Man-made intelligence such a large number of impacts on E-trade on account of the client consistently rulers of the market. Therefore, clients are as indicated by their comfort, give more inclination for internet shopping. Simulated intelligence additionally assumes an indispensable job in it since when buyers do internet-shopping, Voice scan permits clients to look for new things as per their necessities, needs, and wants without utilizing the PC to purchase appropriate and tough products to satisfy their requirements.

1.7.6 APPRAISAL AND FORUM MODERATION

Until this point, each organization has been chosen by Google as one of seven driving AI advancements for social great and has been introduced at Google occasions in California, Bangkok, and Tel-Aviv. Moreover, the investigation into the product will be introduced in the not so distant future at the International Conference of Human-Computer Interface in Florida. In addition, many countries are taking a shot at the sending of Trade Marker for inspectors and furthermore to make it available to the Israeli open for pre-administering choices. Our expectation, at that point, is that other national IP workplaces and government offices get and consider how AI could improve their trademark scan forms for the two inspectors and clients.

1.7.7 MARKETPLACE MODERATION

AI and E-trade all things considered utilized by the businessperson to expanding deals since AI incorporates an individual's insight to be utilized the machines and PC frameworks. Purchasers create new methodologies for assembling the quality item. Man-made realization and E-trade both utilized by agents to make a positive effect on the business world? As of late AI and E-trade complementary to one another. The web-based business has no presence with shoppers; time-to-time purchaser's needs and wants to change yet with the assistance of AI inventive new plans to recognize buyers purchasing conduct and as indicated by buyer's requests made the items.

1.7.8 RETAINING THE CUSTOMERS

Suppose the manufacturer provides quality goods to customers, then the customers should be loyal to the product of accompany. Better services helpful to retain the customers.

1.8 EFFECTIVE UTILIZATION OF ARTIFICIAL INTELLIGENCE (AI) AND E-COMMERCE

Customers have an inclination for online business supplies over the block and enormous firearm stores for one most significant explanation ease zone. While we power mirrors that the web-based business works the equivalent for B2B and B2C organizations, there is an unbounded contrast between purchaser conduct. For instance, for a B2C brand, a customer relationship is present moment contrasted with a B2B brand. A B2B buyer won't go on a shopping binge, not at all like his B2C relating thing and appropriately on. Because of these distinctions, B2B web-based business faces difficulties in the accompanying territories:

1. Delivering excellent customer experience;
2. Supervision cash stream and inventory;
3. Maintaining customer interaction;
4. The customizable item for consumption and modification of its price accordingly.

In the statement of these difficulties, the B2B online business prerequisites to get higher to the level where the web-based business organizations will have the option to satisfy the developing need adequately. Energy about the improvements in the wide open of AI, it would appear that it won't be troublesome. In this chapter, we will take a gander at how AI and ML are satisfying to the eye B2B web-based business.

1.8.1 CUSTOMER SEGMENTATION

In the period of statistics, every companionship has calmed approximately every geographic, demographic, and behavioral measurement and metric. They also have exhaustive obtain data from their company. Despite this, the company is unmoving, not competent to make the most of these facts. Your purchaser personas outward appearance the establishment for the sales and marketing line of attack. AI, you can creep through these data points to improve your buyer personas and fragment your clients in a superior manner. Expressive this will consent to you to pull your strategies for that reason and advance the customer experience.

1.8.2 DELIVER A BETTER CUSTOMER SERVICE

The motor instruction limits of chatbots make them more astute with each management; however, alongside that, chatbots additionally propose the notable beneficial thing of fulfillment. You can experience the past talk logs to discover the most normally posed inquiries and concerns articulated by guests and make your substance around those points. Since content is an important part of any B2B activity, make satisfied around the outcome in the development in deals.

1.8.3 IMPROVED CONTENT CREATION

You can experience the past talk logs to discover the most ordinarily posed inquiries and concerns expressed by guests and make your substance around those themes. Since content is a focal segment of any B2B activity, make satisfied around the outcome in the development in deals.

1.8.4 ENHANCED PERSONALIZATION CAPABILITIES

The openness of proposal motors has, through its uncomplicated Internet business arrangements to customize the shopping experience for clients. In light of the purchaser's past perusing examples and shopping subtleties, the web-based business site can show custom-made item proposals that are related to the purchaser's advantages. Another way personalization helps with dynamic evaluating. Since the typical request, an incentive for a B2B client is higher contrasted with a B2C client; stores can execute vivacious costs to improve the deal.

One all the more methods to do this is in a geological setting. Stores can show items and offer that are increasingly fitting to the spot of the purchaser. While we are discussing personalization, vendors can likewise prescribe guarantees, extra trimmings, and so on utilizing AI.

1.8.5 PRODUCT SEARCH

The regular web-based business search is primarily converted by content; however, with the coming of AI and ML, purchasers will no longer need to depend on content-based detection. Facial appearance, like content to-discourse, can accelerate the purchasing procedure. You can utilize advanced collaborators to check the usability of the item and submit a request. An extra larger than usual advance, in necessities of innovation, around there is picture confirmation. Need to purchase an item, however, don't have a clue about the brand name or its modern condition? Simply click a photograph and transfer it to the application. The AI calculation will carry out the dependability for you by coming back with the most material items.

1.9 ARTIFICIAL INTELLIGENCE (AI) IS TRANSFORMING E-COMMERCE

Every businessman more and more capture the market through producing the quality product and identify the consumer needs, wants, and desires a selling the product. The purchaser is the ruler of the market. At the point when an agent dispatches the new item in the market to present the buyers about their item, embrace the numerous strategies, systems, and innovation for it. Web-based business assumes an imperative job. Since with the assistance of E-trade, effectively advertisers conveyed the item in the hands of buyers and purchasers spare the time simpler to get the item through an online installment. AI and E-trade all in all utilized by the businessperson to expanding deals since AI incorporates an individual's insight to be utilized the machines and PC frameworks. Customers create new methods for assembling the quality item. Man-made reasoning and E-trade both utilized by representatives to make a positive effect on the business world? As of late AI and E-trade complementary to one another. Internet business has no presence with shoppers; time-to-time customer's needs and wants to change yet with the assistance of AI imaginative new plans to recognize purchasers purchasing conduct and as per buyers' requests made the items. Each business needs to gain them increasingly more benefit, benefit dependent on deals of the item, and deals rely upon showcasing technique on the most proficient method to catch the market—the advertising procedure part of E-trade. AI and E-trade consolidate utilized by an organization to arrive at its objective market through different ways like business-to-business, business to customers, and so on.

The investigation likewise underscores the requirement for social changes and skilling and deskilling workforces to make. AI works for the nation to prevail in the AI race; India needs to supportability improves its availability.

Business pioneers a lot of put stock in AI to advance and improved the altruism of the organization moreover.

1.9.1 PERSONAL SHOPPER TO THE EVERYDAY CONSUMER

Organizations are eager to put resources into human capital; however, they need solid focuses. An incredible business needs abilities ought to be requested just as computerized aptitudes. The greatest aptitudes holes distinguished by the relational abilities. Each organization needs to hold steadfast clients in the organization in such a case that your organization gives solid and quality merchandise, as result client free informal exposure of the organization's item.

1.9.2 USED BY RETAILERS

Scanned for a veterinarian in your general locality, and now your social feed has been taken over by promotions for pet items? That is the intensity of AI. Through AI, retailers can find dynamic examples in client propensities and give bespoke shopping encounters. With the intensity of ML, retailers would now be able to try and foresee client decisions even before the client thinks about their necessities. This chapter in Mantra covers in detail how ML can be utilized in foreseeing purchaser conduct interface.

1.9.3 BUYER FACILITATE

Chats bots are as of now utilized in just a few of the web-based business advertise for client help, and this number is on the ascent. Computer-based intelligence is quickly improving chatbots to unravel a lot of questions, yet additionally, give item suggestions and surveys. An individual virtual beautician for your design or a financial master for your fiscal exchanges is a few instances of the AI move in client care.

1.9.4 FORMATION AND REPRESENTATION EDITING

An E-trade store's plan and nature of item pictures assume a basic job in driving deals. Architects have embraced responsive plans to adjust sites to different screen goals throughout the years. In any case, with the utilization of ML, planners are presently endeavoring to customize the site to singular clients. Imagine a store masterminded and showed it to you. With AI, style e-rears are likewise now moving ceaselessly from customarily costly photograph shoots and utilizing AI to take care of the attire photography issue. Picture altering administrations and utilize AI to make completed pictures inside minutes with almost no human intercession by any means.

1.9.5 LOCATE SATISFIED AND LAYOUT

With the assistance of cutting edge EA, retailers can look forward to content advertising results through testing gatherings of potential arrangements. With dynamic learning, more gatherings of arrangements are naturally evolved and tried, in view of the accomplishment of past gatherings. Brand informing, site plan, and design, the suggestion motor is a portion of the key regions that EAs can effectively reshape.

With the headway in AI and better equipment, the online business experience is going to change definitely for both customers and merchants. The inquiry is on the eagerness of e-Retailers to take advantage of this lucky break.

1.10 BASE OF ARTIFICIAL INTELLIGENCE (AI) OF EVERY E-BUSINESS

Man-made brainpower is a limit not identified with just WI-FI, computerized motion pictures, advanced library. Presently it ought to be utilized on a major level. Web-based business assumes a fundamental job. Since with the assistance of E-trade, effectively advertisers conveyed the item in the hands of shoppers and buyers spare the time simpler to get the item through an online installment. Simulated intelligence and E-trade on the whole utilized by the agent to expanding deals since AI incorporates individuals' insight to be utilized the machines and PC frameworks. Shoppers create new systems for assembling the quality item. Man-made perception and E-trade both utilized by specialists to make a positive effect on the business world? As of late AI and E-trade corresponding to one another. The web-based business has no presence with purchasers, time-to-time buyer's needs and wants to change yet with the assistance of AI inventive new plans to recognize shoppers purchasing conduct and as indicated by customer's requests fabricated the items. Each business needs to gain them increasingly more benefit, benefit dependent on deals of the item, and deals rely upon promoting procedure on the most proficient method to catch the market—the promoting system part of E-business. AI and E-trade join utilized by an organization to arrive at its objective market through different ways like business-to-business, business to customers, and so on.

1.10.1 VIRTUAL ASSISTANTS

A remote helper alludes to oneself created and independently employed business. Inventive to new thoughts on the best way to fulfill the customer's necessities, to accomplish more creation increasingly to limit the cost and boost the benefits. On the off chance that somebody gives an ideal shape to his thought, therefore, he not exclusively accomplishes for own self, additionally set a model for everybody to set up his or her business or autonomously employed. Not just offer shape to claim thoughts additionally to be utilized innovation help their business from home. The primary and live case of virtual help is the revelation of occupations and diminished redundancy. It's useful to make the arrangements, make the ticket at the hour of movement, to accomplish the exploration work too.

1.10.2 CHATBOTS

Chatbots very helpful at the time of conversation time by the user to share the views, thoughts through the chatbots. A user conveys their messages in natural languages through the time of testing, editing, to make the websites, also helpful in mobile apps. Chatbots also play a vital role in the classroom to share innovative ideas and opinions. In today's era, chatbots start new trends to share ideas, information, and thoughts.

1.10.3 BIG DATA

All things considered, shockingly, the best model we could discover in this setting is the United States Postal Service's AI-driven location acknowledgment. Despite the fact that it may sound insane, AI is viewed as a break new ground in the field of AI. It was one of the principal associations to begin making significant interests in the innovation of AI in the period of purported massive statistics. Allows simply to consider that for a moment: there are large locations in the United States, and a great deal of web-based shopping to convey. This astounding result gets from the way to established and financed to locate the exact AI system to digitize hand and composed locations. So now, numerous individuals don't have to do this dreary activity any longer. There is currently attention to the tremendous measure of value information being produced each day, together with an understanding that the administration can lead the presentation of AI for the open great. This implies even the most non-innovative driven workplaces can get a handle on that AI isn't as cutting edge as we would might suspect it's nowhere, and it could as of now be utilized for illuminating the present difficulties. One of them is the IP field, and particularly around trademark closeness.

1.11 CONCLUSION

Man-made intelligence and E-trade are aggregately utilized by the agent to intensifying deals since AI incorporates a person's knowledge to be utilized by the machines and frameworks. Through the E-trade business is utilized by retailers, wholesalers, and representatives are as of now utilizing types of AI to more readily comprehend their buyers, create new strategies for assembling the quality item. Man-made understanding and E-business are both utilized by representatives to have a positive effect on the business world; as of late, AI and E-trade harmonizing with one another. The web-based business has no presence with customers, time-to-time buyer's needs and wants to change; however, with the assistance of AI innovative new plans to recognize buyers purchasing conduct and as per buyers, and requests fabricated the items. People have foundation information, intuition, and adaptability no machine can allow. Those stances challenges for AI interpretation in online business human interpreters wouldn't need to manage.

KEYWORDS

- **artificial intelligence**
- **artificial intelligence with marketing**
- **e-commerce**
- **evolutionary algorithms**
- **marketing**
- **World Intellectual Property Organization**

REFERENCES

1. Paul, E. D., & Wolfgang, D., (2019). *The Journal of Artificial Intelligence (AIJ).* Commenced publication in 1970; source-normalized impact per paper (SNIP): 2.800: All articles published after 48 months have unrestricted access and will remain.
2. Ajit, A. (2018). Engineering applications of artificial intelligence, *The International Journal of Intelligent Real-Time Automation.*
3. Amit, K. S., & Manvendra, J. (2018).*Engineering Applications of Artificial Intelligence: A Bibliometric Analysis of 30 Years (1988–2018).* https://doi.org/10.1016/j.engappai.2019.06.010.
4. Jeremy, B., (2019). *Embedding Projection for Targeted Cross-Lingual Sentiment: Model Comparisons and a Real-World Study*. 10.1613/jair.1.11561https://doi.org.
5. Abu, S. B. A., (2005). *A Corpus-Based Approach to Generalizing a Chabot System.* School of Computing, University of Leeds, Leeds, [3] "ELIZA-Wikipedia, the free encyclopedia." [Online]. Available: http://en.wikipedia.org/wiki/ELIZA (accessed on 1 December 2020).
6. Siddhartha, G., et al., (2015). *(IJCSIT) International Journal of Computer Science and Information Technologies, 6*(2), 1483–1485. scholar.googlr.co.in (accessed on 8 March 2020).
7. Loudon, K. C., & Guerin, T. C., (2014). *E-commerce, Business, Technology, Society* (10th edn.). Pearson. ISBN: 978-013-302444-9.
8. Chaudhury, A., & Jean-Pierre, K., (2002). *E-Business and e-commerce Infrastructure.* McGraw-Hill. ISBN: 978-0-07-247875-4.
9. Friedan, J. D., & Roche, S. P. (2006). *E-commerce Positively Impact on Business*, Richmond Journal of Law & Technology, Volume XIII, Issue 2, www.heinonline.org (accessed on 8 March 2020).
10. Graler, A., (2009). *Development and Growth of the Country: A Review about Artificial Intelligence Geography Compass, 2*(3), 771. doi: 10.22111/j.1749-8198.2001.00096.x.
11. Human, P., & Jung, M., (2013). *In-Depth Benchmark of 12 e-commerce Solutions (PDF),* In-Depth Benchmark of 12 e-commerce Solutions (PDF), www.slideshare.net,NBSSystem,https://is.gd/e_commerce (accessed on 8 March 2020).
12. Kessler, M., (2003). *More Shoppers Proceed to Checkout Online.* USA. Today Self-healing structural composite materials, *Composite Part A: Applied Science and Manufacturing, 34*(8), 743–753.
13. Lowry, P. B., Wells, T., Moody, G. D., Humphreys, S., & Kettles, D., (2009). *E-commerce Transactions and Improving Quality Management, 11*(3), 2–44. http://aisel.aisnet.org/cais/vol1/iss3/6 (accessed on 1 December 2020).
14. Kotller, W., (2009). *Marketing Management.* ISBN: 948-21-817-1233-0.
15. Merited, R., (2011). *The AI and e-commerce Background Collective Learning* (p. 741). ISBN: 958-0-324-06100-8.
16. Nissanoff, D., (2016). *Future Shop* (Hardcover edn., p. 24). The Pencert Press. ISBN: 908-1-65420-077-9.
17. Sqrtyquig, (2001). *Consumer Behavior Analysis, 9*(2).
18. Mayhem, R., Siham, Y., Mounia, M., & Hajar, M. *An intelligent Swarm-Based Optimization Technique for oscillatory Stability Assessment in Power System* December 2019, *IAES International Journal of Artificial Intelligence (IJ-AI), 8*(4), pp. 342–351, ISSN: 2252–8938, DOI: 10.11591/ijaj.

19. Kamahi, N. A. M., Musirin, I., Ibrahim, A. A., & Hali, S. A. (2016). "A Review Paper On e-commerce," *Asian Journal of Technology & Management Research*, *6*(1), ISSN: 2249–0892, http://www.ajtmr.com/papers/Vol6Issue1/Vol6Iss1_P3.pdf (accessed on 8 March 2020).
20. Mohammed, B., Salima, T., & Fouzia, O. (2012). "A General View on the E-banking," *43*(13), 62–64. http://www.ipedr.com/vol43/013-ICFME2012-M00033.pdf (accessed on 8 March 2020).
21. https://int.search.myway.com/search/GGmain.jhtml?p2=%5EAFW%5Echr999%5ETTAB03%5E&ptb=5AE30555-3AD4-4007-8069-092A4F995193&n=78671697&ln=en&si=&tpr=hpsb&trs=wtt&brwsid=061BD631-2130-401E-8D4-B-DDC6F4E45DCD&searchfor=www.wikipedia.com+related+to+e-commerce&st=tab (accessed on 8 March 2020)..
22. www.wikipedia.org (accessed on 1 December 2020).
23. https://www.abacademies.org/articles/ecommerce-and-the-newspaper-industry-determinants-to-firstmovership-6575.html (accessed on 8 March 2020). (accessed on 1 December 2020).

CHAPTER 2

ARTIFICIAL INTELLIGENCE: A NEW DRIVER FOR MANAGING CUSTOMERS IN E-COMMERCE SMARTLY

SHARMISTHA DEY

Maulana Abul Kalam Azad University of Technology (WBUT), Kolkata, West Bengal, India, E-mail: papri.dey@gmail.com

ABSTRACT

e-commerce is one such industry that preferred to stay back from the use of AI applications in the initial stages but finally, it has begun its journey towards the recent AI era. E-commerce also has faced a huge paradigm shift using AI-especially machine intelligence (MI), data analytics, and business intelligence.

Online businesses nowadays are facing highly notable competition, and data is one of the most valuable resources for the flourishing and nourishing of any business. Presently, most of the businesses are app-based, and they share a virtual marketplace. Therefore, getting insights from the data and generating knowledge and wisdom from the information is the next main aim for getting a better customer experience, better decision-making, recommender system, or setting the dynamic pricing intelligently according to customers' buying patterns and characteristics. This proposed chapter mainly focuses on the area where different AI techniques are used to improve customer experience in e-commerce.

For better customer experience and getting a competitive advantage to create a new marketplace, e-commerce industries are now utilizing data analytics and MI, which in turn, a part of artificial intelligence (AI). Though the perception of AI is not new, widespread application of AI is emerging since the last couple of years, and this new era is the era of AI.

The proposed chapter says about AI and its types, different AI techniques that are being applied by several e-commerce industries, and the present trends along with the future direction of AI for the e-commerce industry.

2.1 INTRODUCTION

e-commerce or online commerce has witnessed a drastic change in the last few years after being emerged with the power of AI. Though initially, the e-commerce industry was on the back foot and quite reluctant to adopting artificial intelligence (AI). However, the adaption of technologies like machine learning (ML), data analytics, and robotics have made this industry flexible and more approachable for gaining a competitive advantage in this highly competitive business world.

AI has created a paradigm shift in almost all industries-from manufacturing to marketing, from biomedical to healthcare, or from education to entertainment. The concept of AI is not at all a new one since the last couple of years, AI has reached a new milestone with the evolution of AI technologies like ML, data analytics, cognitive analytics, etc. AI has become the best choice for researchers and creating several innovative applications in almost all fields of science, technology, or management [1–5].

AI is the science in which we can understand how a machine can think, understand, or interprets any real-world object. In AI, the computer can follow the real being and act according to their behavior.

e-commerce, on the other side, very essential for today's business scenario. E-commerce or Electronic Commerce is the technique using which the vendors can sell their products, concepts, or services online, and also the buyers can get a 24×7 available market, sitting at their home.

e-commerce businesses have late in the world of AI, but the highly competitive marketplace in e-commerce has witnessed a paradigm shift due to the innovative applications of AI. Nowadays, online businesses are using AI applications for enhancing their customer, having better customer experience, or for dynamics price-fixing.

2.2 ARTIFICIAL INTELLIGENCE (AI)

AI has created a revolution in almost all the fields of science, technology, and innovation is the intelligence in the machine, where the machine has consciousness or understanding, and it can take decision-getting insight from the training data.

2.2.1 TYPES OF ARTIFICIAL INTELLIGENCE (AI)

Mostly by AI, we mean such an application where the machine can learn from some training data and perform a limited task after learning from the supervised or unsupervised training data, but other than this, a machine can also be an expert in self-learning and autonomous. Depending on learning capacity, expertise level, and some other factors, the AI systems have been categorized into the following different categories [7–9]:

1. **Artificial Narrow Intelligence or ANI:** These AI applications are such an AI application that can perform some specific and limited applications using human-like capabilities. They perform only the assigned work or only some specific program according to the previously fixed routine. Apple's face recognition technology can detect the face, Google Assistant, or Google Photo app, etc. Different types of Chat-Bots also can be a good example of Artificial Narrow Intelligence. The marketing area also has ANI applications. Agents with artificial narrow intelligence are designed in such a way that they can solve some limited problems with a lot of success, and those agents have set the customer experience to a very high altitude. Most deep learning and intelligent machine applications belong to this category. Alexa or SIRI is an example of such intelligent agents having artificial narrow intelligence.

   The use of artificial narrow intelligence in the development of business is a widely accepted technique nowadays to get a competitive advantage in this worldwide virtual market.
2. **Artificial General Intelligence (AGI):** It is such an AI application where the machine can learn by itself. Here, the machine can learn from the training data, observe, and comprehend by an AI agent as similar as a human. This artificial general intelligence (AGI) will be self-learning and self-directed. AGI will make an AI system as if it is making the decisions as a self-dependent human being. Some researchers have termed AGI as strong AI also, as the machines are smarter enough and proficient to decide the consciousness.

   Today more priority is being given to the development of the AGI system. Turing test, coffee test, Robot challenge tests are an example of some tests that have been performed in years to test AGI skills in the machine. The main aim of AGI development is trying to find whether a machine has consciousness or not, compare to a human being.

3. **Artificial Super Intelligence (ASI):** It is to some extent like AGI. However, it is yet to be implemented in reality; till now, it is more a concept. Scientists are working very fast in this area; they are trying to work in developing super AI. Nowadays, some systems are even more perfectly can do work than a human being. It can be able to overpower human performance [5, 6]. ASI is the result of the development of cognitive ability in the system, making the machine more superior to humans [7]. However, scientists are a little bit afraid of developing ASI as it may be unstoppable. The fear is someday, the exploration of ASI can't be stopped by humans, and it may raise a signal of danger for the human race itself. Machines are becoming smarter and capable enough to produce a replica of the intelligent machine; still, these fields have a huge scope of the invention, especially in human-computer interaction, cognitive ability, augmented analytics, etc. (Figure 2.1).

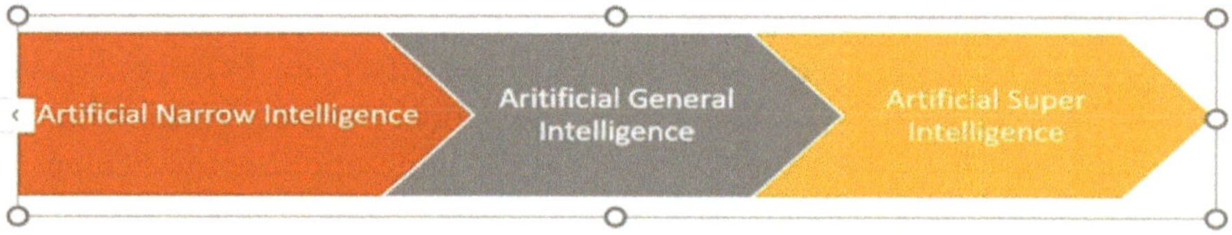

FIGURE 2.1 Artificial intelligence types.

2.3 E-COMMERCE AND ITS TYPES AND APPLICATION

e-commerce or electronic commerce is an area where the buying, selling, or maintaining a product or service is being done using the internet. With the advancement of technology, e-commerce industries started to flourish. The high-speed internet with a low-cost availability has made it possible to reach the network even in the remote corner nowadays, and this is one very prime factor for e-commerce being popular day by day.

e-commerce is a superset of online business where, along with buying or selling goods online, the transfer of money and data to execute those transactions is being done with the help of the internet. The user or customer can send a request using a browser application from the client-side, and it goes to the corresponding server; the server searches from its database and sends an acknowledgment as a response. Therefore, this service is a multilayer architecture.

The e-commerce industry is manifested by the use of AI nowadays. There are various sub-areas where AI has created its impact. Using several ML algorithms like classification, support vector machine, regression analysis, or some other time series analysis, the price forecasting done in an e-commerce application, using smart algorithms, and robotics, the concept of warehouse management (WM) has been changed. Nowadays, many e-commerce companies like eBay, Amazon, OYO, Goibibo are using several predictive analytics techniques for better customer service. CRM is also an important area for AI implementation. For managing customers smartly and intelligently, nowadays the companies are using NLP based chatbot and intelligent agents like Alexa, Siri, Kisha, etc. All major CRM solutions are adopting AI for better service; they are buying AI startups (Figure 2.2):

1. **Ubiquity Property:** The services for e-commerce is available everywhere. A user sitting from anywhere in the world can book any product from the other corner of the world using e-commerce. An example is shopping through Amazon or alibaba.com [9].
2. **Global Reach:** Your business will have a worldwide reach. The entire number of target users and customers for e-commerce is distributed around the world. There is no geographical restriction. In addition, language is not a barrier, as nowadays, most of the sites are multilingual [9].
3. **Universal Standards:** The standards in e-commerce have been decided by a common body containing members from every corner of the world. The price of certain particular products you see on the website is fair and standard, though it may vary from site to site as per the discount available.
4. **Interactivity:** 21st century's business is interactive; the world is in our grip. So bi-directional communication among businesses and consumers is allowed.
5. **Information Density:** The information density has been improved after the explosion of the internet. It has enriched the total capacity of the market, buyers' reach, etc. The e-commerce technology condensed collection of information, processing cost, and storage communication. At the same time, it also reduces accuracy and timeliness [9].

6. **Personalization:** e-commerce permits personalized options for customers. Businesses can be familiar with one's name, a user's list of interests or product choices, and past purchase patterns, and according to those, the customer may be given options. The technology allows the consumers for custom behavior. The vendors can display their products or services by studying the users' preferences or previous buying patterns [9].

FIGURE 2.2 Features of e-commerce.

2.3.1 *TYPES OF E-COMMERCE APPLICATION*

e-commerce businesses mainly fall in the following category:

- Business-to-business (B2B);
- Business-to-customer (B2C);
- Customer-to-business (C2B);
- Consumer-to-consumer (C2C) e-commerce, etc.

According to business buyers and consumers, the type of e-commerce applications may be illustrated using the following categories:

1. **Business-to-Business (B2B e-commerce):** This is the transaction between two vendors. Some companies are running online commercial transactions with each other, without the involvement of any consumer. The pillars of business, like manufacturers, wholesalers or retailers, etc., are involved in this. Generally, it is used to increase the efficiency and effectiveness of a company's sales performance with online support.

   B2B e-commerce is one of the fastest-growing e-commerce segments, which could touch a large amount of transaction in the next few years. Their applications are in electronics, shipping products, motor parts, petrochemicals, paper, food agriculture, etc., are the area where B2B e-commerce has its applications.

   The key technologies behind B2B e-commerce are:

   i. **Electronic Data Interchange (EDI):** It is nothing but an inter-organizational exchange of data or information about organizational documents that are in a structured format and machine producible.
   ii. **Internet:** Without the internet, we can't proceed almost in any sector today. It is a huge network, connecting other persons from the world or outside my network. We can say the networks of a network that can connect the whole world. Here the client sends the request from the browser application, and the request goes to the corresponding server, and it returns with the response.
   iii. **Intranet:** It is a dedicated network within an organization. It is more secure and reliable than the internet.
   iv. **Extranet:** It is a wide area network, connecting the outer world having inadequate access to a certain area of an enterprise network.
   v. **Back-End Information System:** These are the database server used to organize the business data.

   B2B e-commerce follows the following models:

   a. **Supplier Oriented Marketplace:** Here a common online business platform provided by the supplier for promoting their sales promotion. It is a common marketplace for the suppliers and customers where the suppliers promote their sales products.

b. **Buyer Oriented Marketplace:** It is the marketplace launched by a buyer where the buyers can bid their products online.
c. **Intermediary Marketplace:** It is a place where a buyer and seller can meet through a third party. Here an intermediate company directs the whole marketplace where the buyer and sellers meet and perform; for example, Buyswings, Assurant, Freund, Human Solution, etc.

2. **Business to Customer or Consumer (B2C e-commerce):** Interactions between business-to-customer and business-to-consumer. Here, the buyer company trades its products or services to the consumer without the direct intervention of a third party. Consumers can browse the websites of the vendors and survey or compare their products, pictures; they also can read product reviews. After that, if they find it suitable, the consumer can place their orders, and the company delivers the goods straight to them. This B2C model was popular during the late 1990s.

   Any business involved in B2C business must maintain a good relationship with their customers and provide the best support possible to the customers as the competition is very high.

The e-commerce businesses are of the following basic five types:

i. **Direct Sellers:** This is one of the most popular models where users purchase products from online vendors. These may comprise the producers or even small businesses. Consumers may purchase products from online stores collecting products from several manufacturers.
ii. **Online Intermediaries:** This model is basically for the third party or the intermediaries who don't any product or service but as a third party; they run their services for the vendors and put buyers and sellers together. Different online travel websites like Expedia, Trivago, and Etsy are an example of this category.
iii. **Advertising-Based B2C:** This new e-commerce model allows permitted content to appeal to visitors to a website. Those visitors, in turn, display digital advertisements. A huge volume of web traffic is used for the purpose of advertising. These sites help in the marketing or promotion of any product or service. There are different media sites for this, e.g., Huffington Post.
iv. **Community-Based:** Nowadays, businesses or marketing activities may also be done using social media. Digital marketing through

social media is quite popular nowadays. Facebook, Snapchat, or LinkedIn, build connected communities based on users' common interests, which could help the marketers and advertisers for promoting their products straight to the consumers. Such websites usually target their promotional content based on the particular users' demographic data and the basis of their geographical location.

v. **Fee-Based:** Some websites charge a reasonable amount from the users or the consumers for providing their service or offering their products. The site may also offer a free limited context. Netflix and Amazon Prime are an example of this. Example of B2C e-commerce site: Amazon, Flipkart, Jabong, etc.

3. **Consumer-to-Consumer e-commerce (C2C e-commerce):** This e-commerce transaction leads to interaction among consumers itself, without the interference of any buyer company. No vendor or business is involved. Such kind of sites could help the users to trade their household personal products or assets unswervingly to another interested consumer, without the intervention of a business. Usually, any household products, electronic gadgets, or required things may be transferred through such kinds of E-Business sites. Example: OLX, Quikr, etc., are following such a model.
4. **Consumer-to-Business e-commerce (C2B e-commerce):** The reverse concept of B2C e-commerce. Here, the user provides a product or some service to the vendor. Consider the situation of an IT out worker giving an online demo of his work and retails his developed software to a company. This is an example of the consumer-to-business (C2B) transaction. Example: photolia.com.
5. **Government to Business e-commerce:** It is also known as Government to Administration e-commerce, where the government announces notice or procurement or gives online tender for the businesses. Any governmental site or government E-auction comes under this category. Example: Any government site for E-tender or procurement, for example, notices for online tender from the railway.
6. **Government to Citizen:** If the government wants to share any message with the citizen or they want to pursue any process for the citizen that comes under this category. The birth registration site for the citizen or Government job sites, etc., which reduces the normal application time for the citizen (Figure 2.3).

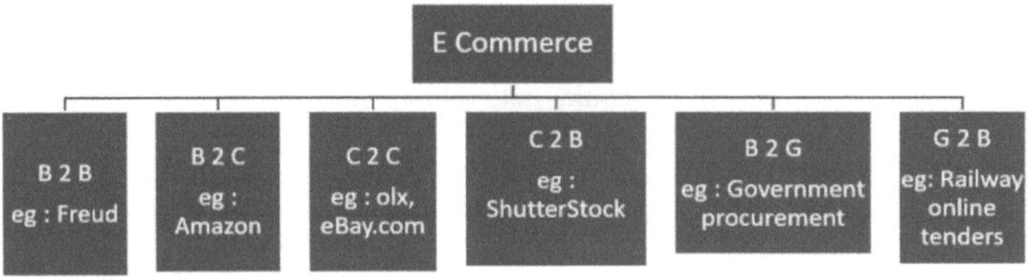

FIGURE 2.3 Type of e-commerce businesses.

2.4 WHY E-COMMERCE INDUSTRY WAS INITIALLY RELUCTANT TO JOIN THE AI CAMP?

The e-commerce sites were initially was reluctant to adopt AI, as initially, the market was not ready. However, there was a gap between the preferences of the customers and the supply or the distribution of promotion according to customers' choice. No concept of visual searching was there. The personalization in social media like Facebook or the personalization in the e-commerce sector both can't be avoided. Inventory management is the biggest asset of the e-commerce sector. Though security and reliability were the major concern for the reluctance of the e-commerce industry to the setback, later on, the emergence of AI has changed this industry into a completely new lookup. Now e-commerce industries are adapting AI algorithms to change the shape of the business of the new age. From hotel industries to retail or from their promotional aspect to customer service, everywhere AI has created a new vision.

The major limitations of using AI in e-commerce may be considered through the following points:

1. **Data Availability:** One of the main barriers to the implementation of AI is data or information availability. Data is unpredictable, and no guaranty of the quality of data, those present tasks for businesses taking care of generating value from AI at scale. To overawe such problems, the organization should follow a clear strategy at the very beginning for sourcing the data in which an AI solution will be needed.
2. **Shortage of Skilled Person:** Another key problem area to adopt AI is the shortage of skills or shortage of AI experts, as availability of properly skilled staff with the experience and training necessary to successfully arrange or operate AI solutions. The lack of skilled data

scientists and proper infrastructure was also a matter of concern for the adaptation of AI in e-commerce.

3. **Cost-Cost:** It is another key point for discussion for the reluctance of the e-commerce industry to adopt AI technologies. In the absence of in-house skills, AI is not suitable. Therefore, the products often have to subcontract, which incurs the contests of cost and maintenance appears. Due to high complexity, smart technologies are highly expensive, generating further costs for fixing and continuing charge for maintenance. The computational cost of training data models considered to be an additional expense.
4. **Security:** It is one major concerning area for adaptation.

2.5 AI AND E-COMMERCE REVOLUTION: HOW AI HAS CHANGED THE E-COMMERCE MARKET

After the revolution of AI, specifically, the emergence of ML and data analytics, has made a huge paradigm shift in the e-commerce market. Nowadays, data is the main resource for online companies, the customer is their asset, and by using personalized service or recommendation, the online sites are doing a lot of businesses and gaining competitive advantages.

There are many areas in e-commerce where researchers have given the limelight. AI has made its footprint in every angle of e-commerce-from supply chain management (SCM) to customer relationship management (CRM), or in recommender system or dynamic pricing, or it may be the newly trending E-negotiation or E-Auction.

In the following section, some of the related studies on the impact of AI in e-commerce have been discussed.

2.5.1 RELATIVE RESEARCHES ON THE IMPACT OF AI IN E-COMMERCE APPLICATIONS

Businesses like e-commerce have entered into the AI era lately, but to gain the advantage of a highly competitive market, data is the main resource nowadays, and businesses are becoming sharper, smarter by adapting new cutting-edge technology for growing their businesses. The whole business model is being shifted towards a new paradigm with the help of AI.

The authors Kumar and Trakru [18], in their paper "*The Colossal Impact of Artificial Intelligence in e-commerce*," have shown about complete impact of AI on e-commerce applications. This chapter has shown an overall discussion regarding the impact of AI on e-commerce and its statistical impact. This is not a very strong chapter but covers the general discussion about the impact of AI and ML on e-commerce [18]. What are the several components of AI, and how they have created an impact in e-commerce? These are the main discussion area; however, this chapter is more a general approach based. However, AI is changing the way of online transactions or changing the whole e-commerce world. The impact of the ML technique is up to a great extent in the case of e-commerce applications. ML may be termed as a subset of AI.

In their work, Mecham et al. have explored adaptivity and recommendation methods and implemented these methods for an e-commerce web-shopping site. The authors have used python framework technology for their implementation, using Python Web MVC architecture [*Model View Controller Architecture:* It is known as model view controller architecture. Here model means data or business logic, the view is a user interface, and the controller is a request handler. In Python, Django uses the MVC architecture (Figure 2.4). Django is a very logically structured framework in python used for MVC architecture.] [22]. This work has presented AI decisions through adaptive interfaces. The authors have a mixed machine-learning algorithm and applied it over the adaptive web-based framework using PHP, Django, and angular JS environments. Several AI applications may be seen in the case of an e-commerce transaction.

Srinivasan et al. [1] has discussed several AI applications in their chapter. They have communicated about ML, artificial neural network, data analytics, etc., technologies which have created a great influence for developing.

Researcher Jasim [2], have demonstrated the DHCP or dynamic host configuration protocol. The researchers argue that the LAN can be made event-driven, and they also have discussed how an online system, Expert system, and AI affect research products. They have also discussed the need for XML, which is designed to transport and store data. This chapter is security-oriented to e-commerce, which is much more important in distributed architecture through the web. The author has emphasized the aspects of security but has not applied AI techniques to make the system more intelligent and smart, which could be a better approach [2].

One of the main facets of AI application in e-commerce is data analytics or, more specifically, predictive analytics. It can be done in many ways like data analytics, time series forecasting or opinion mining, or big data analytics. The opinion of the online customers incurs a lot of value for the sales and promotions of a product, so technique using opinion mining is one of the popular techniques for the extraction of insights from data. The paper *"Real-time Sentiment Analysis on e-commerce Application"* says about sentiment analysis, where Support Vector Machine or SVM algorithm has been applied to get a better impact in sentiment analysis [14]. The authors have used a support vector machine technique to design a model for an e-commerce application. The main aim of this chapter is to build a real-time sentiment analysis on product reviews given by different customers, and based on experience, the conclusion can be drawn. The dataset used for this chapter has been collected from the Amazon site in JSON format, and this has been collected from several product reviews made by the customers on the Amazon site. The authors have used a supervised learning model and classification method. The proposed model can judge a positive or negative response.

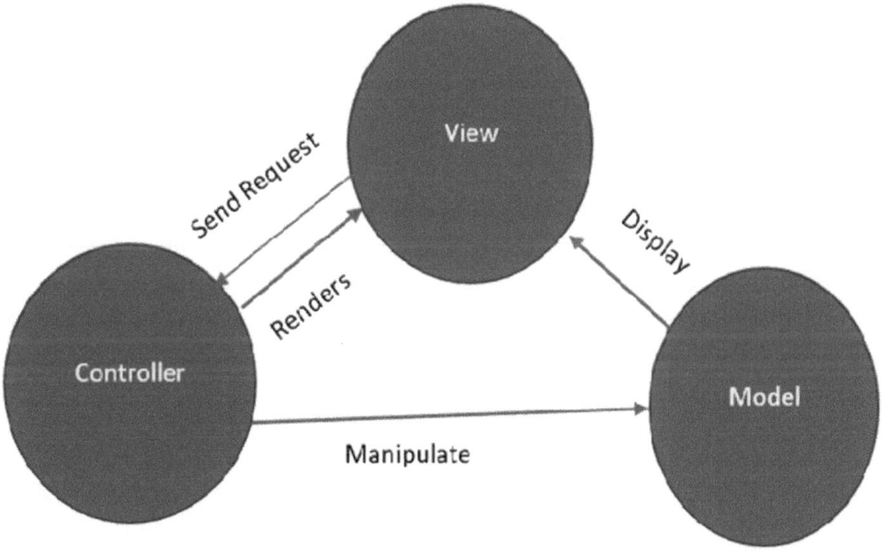

FIGURE 2.4 MVC architecture.

The specialty of their model is to exhibit the real-time predictive analytics of the product review. According to the authors, sentiment analysis can be used for any online shopping applications like Amazon or Flipkart, etc., and with the help of AI applications, the customers can get an educated opinion.

Today the sites like Amazon or Flipkart are using this sentiment analysis to judge the review or remove biases.

In recent years, online transactions have acquired popularity, and with the rise of online transactions, the cases of fraud have also increased. Saji and Panchal have discussed this in their chapter [15]. They have discussed fraud detection by a technique called "adaptive neuro-fuzzy inference system," which is a combination of artificial neural networks and fuzzy inference. In their chapter, the authors have discussed how different online fraud occurs. According to them, the major concern of fraud detection is about to detect the fraud within a very small time-frame available. The authors have proposed the Adaptive Neuro-Fuzzy approach for fraud detection. The work is demonstrated as an analytical technique of computational intelligence. However, the main challenges were an available primary dataset. In this area, some other researchers are also there.

Another aspect of AI application in e-commerce application is in Recommender System. A recommender system, or sometimes known as a "*recommendation system*," which is one type of information filtering system, seeks to predict the rating or preferences for a user [17]. Huang, Chung, and Chen, in their chapter, have proposed a framework for a graph-based recommender system. This chapter is an old one still efficient to describe the effect of AI in e-commerce. Here the authors have used a Chinese online bookstore as their testbed. The authors have discussed several efforts that have been made to improve the quality of the recommender system. Most of the recommender system uses the collaborative filtering effect to build a recommender system, according to the authors. The authors proposed a two-layer graph model for the recommendation in their system-one is a product layer, which is content-based and the other is the customer layer, which is demographically based. They have considered content-based, collaborative filtering, and a hybrid approach.

In another paper entitled *"An Adaptive e-commerce Application using Web Framework Technology and Machine Learning,"* the authors Meacham et al. have proposed a recommendation framework [22] using ML techniques. They have shown through use cases how their system works and increases productivity. They have improved their recommendation system using machine-learning techniques. The authors have performed a content-based recommendation system, the first approach is to develop with the help of a cookie-based system, and the second approach is developing the same with the help of python library scikit using a machine-learning algorithm. However, the authors are silent about the trustability of the system, and it could be improved with some other machine intelligence (MI) algorithm also.

Other papers show work on the recommender system using MI or data analytics. Pavithra et al. [20] have discussed the effect of Big Data analytics in e-commerce. Big data is a huge amount of data with multiple dimensions like velocity, variability, variety. When it is the matter of e-commerce, the question of security is the priority. There is a difference between data processing of traditional data and big data. In this chapter, the authors predicted the users' behavior for online advertisements on predictive data. They used data retrieved from several social media like Facebook or Twitter. This is a survey chapter, and no specific research proposal has been introduced in this chapter. They used three types of analytics-social media analytics, predictive analytics, and mobile analytics. Graph analytics and ML. This will gain customer satisfaction and bring more sales figures [20].

Rao, Liu, and Zing have discussed applying ML techniques to discourse the pros and cons of Big Data mining. According to the authors of this chapter, ML can solve the problem by acquiring prior knowledge [21]. They have used the Naive Bayesian technique of ML for precision marketing. The result of this chapter has shown that machine-learning can not only be able to improve the performance of big data mining, but also it can achieve precise marketing. Today the marketing field has been changed from product-centric to customer-centric. They have used the clustering technique for a further accurate result.

Another aspect of AI application in e-commerce is a negotiation model. Electronic negotiation is one of the popular topics nowadays. Shaobin and Aihua [19] have shown an electronic negotiation model for e-commerce in their chapter [19]. The authors have proposed an automated negotiation model (ANN) for electronic negotiation, which is a learning model using case-based reasoning (CBR) and a neural network. The authors have discussed multi attributive theory (MAUT), which focuses on the evaluation of choices and outcomes of a decision problem. The authors in this chapter have proposed a CBR and neural network-based negotiation process model where the sellers receive the recommendation of the buyer, and then he tests and also he has a chance to utilize the utility function (a function proposed by the authors) to evaluate the proposal. The model is a put forward model for electronic negotiation. However, ANN being a black box model, there is no chance of checking the plausibility of the model, and it is a drawback of this model.

2.5.2 SOME PRACTICAL EXAMPLES OF ONLINE E-COMMERCE SITES WHICH ARE USING E-COMMERCE APPLICATIONS

AI is taking over the world in the last few years. Especially 2018–2019 may be known as such years when the curve of AI will touch the milestone for reaching the online business to a very great extent.

Some e-commerce sites are using AI like ML algorithms or data analytics to get a higher competitive advantage innovatively. Mainly the use of AI technologies like ML, robotics, etc., can be used in various ways to improve e-commerce sites like building an intelligent agent or smart recommendation engine or building a chatbot [23].

AI has made a huge paradigm shift in e-commerce in various ways-from SCM to WM, or from CRM to value chain management (VCM), everywhere the use of AI has innovatively changed the road of e-commerce.

There are some areas where AI has booted up and has made a paradigm shift until the last few years. The areas have been discussed in the following section:

1. **Supply Chain Management (SCM):** The supply chain is a vast area for creating the impact of AI. It is the network of facilities for including several entities in the network. In "*Applications of Machine Learning Techniques in Supply Chain Optimization,*" the authors have discussed the gap between the real demand of the customers and the anticipation made by the vendors [25]. The use of several machine-learning algorithms to make smart logistics solutions, the use of AI and robotics for WM, the use of several ML algorithms, and chatbot for procurement are the current trends of e-commerce applications. The use of robots for better WM and use of chatbot and intelligent agents for better CRM is nowadays trends for the e-commerce industry. The authors in the previously mentioned chapters have discussed various applications running in SCM as well as the future trends on a theoretical aspect.
2. **Smart Logistics:** It is a new buzzword in the supply chain industry of e-commerce. With the magnificent power of AI, logistics handling in the case of e-commerce, product delivery has touched a new height. With the help of IoT enabled d devices and service automation, the delivery of the ordered product to the customers has become easier. The pioneer of online businesses like Alibaba is using smart address mapping technique to improve its logistics, which reduce their shipping time and incurs fewer logistics cost also. Intelligent asset

tracking tools directed by various sensors can offer insight at the individual step of the shipment. The crop health-monitoring tool for tracking moisture or heat at the time of their distribution help to keep track of the vegetable or poultry products' conditions. The use of IoT in logistics will regulate inventory and mobile assets efficiently and innovatively. Now some innovative, sensor-driven tools are providing insights. The use of smart logistics in e-commerce can provide better and intelligent infrastructure, better synchronization of the dynamic logistics system, and better economy management by constantly updated scenario analysis. Today, many companies like Alibaba, Amazon is using smart logistics, with drone-based delivery and intelligent inventory tracking system for better shipment, and it has created a new signature in the realm of e-commerce logistics.

3. **Customer Relationship Management (CRM):** It is a prime factor of any e-commerce business, specifically thinking about the competitive environment, where the efforts given to better customer satisfaction may generate more numbers in the achieve list.

Automation in CRM has created a revolution. Nowadays, some programming solutions have been designed, which can take care of back up administrative jobs. Companies are mixing up AI, and CRM for predictive lead generation, automatic trend plotting for stock markets; AI is also being used for sales process optimization. AI mixing with CRM is helping in upselling, cross-selling of products, demand forecasting, price optimization, etc.

Nowadays, different chatbots are being used using NLP and other intelligence approaches to make customer service much better. The chatbots like Alexa from Amazon, Siri from Apple, or Ali Assistant like Alibaba are some marvelous examples of AI applications in e-commerce for better customer satisfaction.

Another prospective area of CRM is a recommender system. There are several recommendation algorithms in the recommender system, and using this personalized recommendation is possible to send to different customers according to their behavior or buying pattern. The best example is Netflix, which has made its movie recommendation purely on a customer choice basis, and for that, they have been used the help of data analysis and collected insight from those data.

With the help of different ML algorithms, data analytics, forecasting methods, the researchers have been able to anticipate customers' needs, faster respond to customer queries, and monitoring of customer status. Table 2.1 shows some companies and how they are innovatively using AI [21–24].

TABLE 2.1 How Several e-commerce Sites are Using AI in an Innovative Way

Name of the Company	Product Name	Which technique they are Used
Amazon	Alexa, personalized recommendation engine, etc.	Alexa is a voice-based intelligent chatbot developed by Amazon; it helps drive those algorithms helped for the marketing strategy set by Amazon. Amazon has also developed a personalized service based smart recommendation engine, which draws products for a customer based on their past behaviors and buying patterns.
Alibaba	AI Assistant-Tmall Genie and Ali Assistant, use AI for searching most efficient delivery root, smart logistics.	Alibaba's customer service NLP based chatbot manages 95% of customer queries, which are both written and spoken assistant. Their smart logistics help in a 10% reduction in vehicle use and a 30% reduction in the travel budget.
JD.com	Robots for warehouse management and delivery	A Beijing based company that uses robots for their warehouse management. After applying this technique, their manpower requirement has been reduced, and work delivery timing has been reduced, work efficiency increased, and hence the profit margin also increased.
eBay.com	ShopBot	e-Bay's AI chatbot helps to maintain customer interest and to get a competitive edge. They have developed this using natural language processing or NLP [*Natural Language Processing (NLP)***:** It is a subcomponent of computer linguistics, and one application of AI. NLP usually involves speech recognition, handwriting recognition, recognition of natural language, and natural language generation. The practice of NLP started in the 1950s era. For developing many speech recognition problems, chatbots, etc., NLP is being used. Recent research is focusing on supervised and unsupervised learning algorithms. There are two types of NLP: rule-based NLP and statistical NLP. With the emergence of statistical algorithms in the 1980s era, the statistical NLP has been evolved more.]
ASOS	Voice recognition system, image recognition technology	This international fashion retailer has invested in AI for influencing their buyers' behavior.

TABLE 2.1 *(Continued)*

Name of the Company	Product Name	Which technique they are Used
		They are also using Image Recognition technology for a visual search, allowing ASOS photo and finding a matching one or similar cloth for the customer.
Rakuten	For predicting customers' behavioral pattern and sales forecasting, the FitMe app	This Japan-based e-commerce giant has used AI for predicting its customers' behavioral patterns. They have opened an institute for this. Their FitMe app uses image recognition technology and incurs a higher degree of customer satisfaction and sales productivity.
Netflix	For customer choice based predictive service	Netflix uses analytics to understand their customers' choice of movie.
Starbucks	AI Chabot	Starbucks' chatbot, with the help of AI, leads to higher customer satisfaction. It helps the customer to place the order, also helps the customers to tell the preparation time and payment amount.
Pinterest	Automatic back up process	They are using image recognition technology for doing this, which can reduce the workload.
Rolls Royce	Autonomous, partnership with Google	Using AI algorithms, it has been made which can sense what is around them and can predict the danger, and according to them, they pose the ship. This makes shipment faster.
Salesforce	Salesforce Einstein,	Salesforce Einstein is the First comprehensive AI for CRM made by applying deep learning, predictive analytics, and data mining. Rolustech is a SugarCRM made by salesforce.
Zoho	Zia	It is a conversational AI Assistant, a voice assistant.

KEYWORDS

- **artificial general intelligence**
- **chatbot**
- **machine learning**
- **sentiment analysis**
- **smart logistics**
- **supply chain**

REFERENCES

1. Srinivasan, J., Deepika, R., Sreedevi, & Jone, A., (2018). Artificial intelligence of e-commerce platform. *Asia Pacific Journal of Marketing and Management, 7*(8), 1–11.
2. Jasim, A. Y., (2011). An improvement of E-systems over networks via the use of artificial intelligence. *CSRID Journal, 10*(1), 1–8. Hal.
3. URL: https://en.wikipedia.org/wiki/Timeline_of_artificial_intelligence Kanpur, India, (accessed on 1 December 2020).
4. https://wikipedia.org/wiki/history_of_artificial_intelligence (accessed on 1 December 2020). Kanpur, India.
5. Liang, C. J., (2016). *R in Finance: Introduction to R and Its Application in Finance,* The Finance and Financial Planning HDR Research Methodology Workshops at: Griffith University
6. Willium, G. F., (2009). *Financial Engineering Analytics: A Practice Manual Using R* (1st edn.). IEEE Computer Society.
7. Rebecca, R., (2019). *A Complete History of Artificial Intelligence.* URL: https://learn.g2.com/history-of-artificial-intelligence#ai-3 (accessed on 1 December 2020). Kanpur, India, GMT.
8. Naveen, J., (2017). *Seven Types of Artificial Intelligence.* URL: https://www.forbes.com/sites/cognitiveworld/2019/06/19/7-types-of-artificial-intelligence (accessed on 1 December 2020). Kanpur, India, GMT.
9. Luke, M., (2013). *What is AGI?* URL: https://intelligence.org/2013/08/11/what-is-agi/ (accessed on 1 December 2020). Kanpur, India, GMT.
10. Akshay, A., (2019). *Who's Afraid of Artificial Super Intelligence*. URL: https://www.analyticsindiamag.com/whos-afraid-of-artificial-super-intelligence/ (accessed on 1 December 2020). GMT.
11. UKEssays, (2018). *The Seven Unique Features of e-commerce Essays*. URL: https://www.ukessays.com/essays/information-technology/the-seven-unique-features-of-e-commerce-essay.php (accessed on 1 December 2020). Kolkata, India.
12. Lee, K. J., (2000). *Artificial Intelligence Applications in Electronic Commerce* (p. 4). PRICAI 2000, LNAI 1886, © Springer-Verlag Berlin Heidelberg.

13. https://www.tutorialspoint.com/e_commerce/e_commerce_b2b_mode.htm (accessed on 1 December 2020). Kolkata, India.
14. Jabbar J., et al., (2019). Real-time sentiment analysis on e-commerce application. *2019 IEEE 16th International Conference on Networking, Sensing and Control* (pp. 391–396). Banff, Alberta, Canada.
15. Saji, J., & Panchal, D., (2017). Improved fraud detection in e-commerce transactions. *IEEE International Conference on Communication Systems, Computing and IT Applications (CSCITA).*
16. Iskiye, F. O., Folajimi, Y. O., & Ojokoh, B. A., (2015). Recommendation systems: Principles, methods, and evaluation. *Egyptian Informatics Journal, 16*, 261–273.
17. Huang, J., Chung, W., & Chen, H., (2004). A graph model for e-commerce recommender systems. *Journal of the American Society for Information Science and Technology, 55*(3), 259–274.
18. Kumar, T., & Trakru, M., (2019). The colossal impact of artificial intelligence in e-commerce. *International Research Journal of Engineering and Technology, 6*(5), 570–572.
19. Dong, S., & Li, A., (2010). Negotiation model based on artificial intelligence in the e-commerce. *IEEE Third International Symposium on Information Science and Engineering,* (pp. 193–196).
20. Pavithra, B., et al., (2016). The study of big data analytics in e-commerce. *International Journal of Advanced Research in Computer and Communication Engineering, 5*(2), 126–131.
21. Rao, H. K., Liu, A. P., & Zing, Z., (2018). Research on personalized referral service and big data mining for e-commerce with machine learning. *4th International Conference on Computer and Technology Applications* (pp. 35–38).
22. Meacham, S., et al., (2018). *An Adaptive e-commerce Application Using Web Framework Technology and Machine Learning.*
23. Novoselstseva, E., (2018). *Artificial Intelligence in e-commerce: Benefits, Statistics, Facts, Use Cases, and Case Studies.* URL: https://apiumhub.com/tech-blog-barcelona/artificial-intelligence-ecommerce/ (accessed on 1 December 2020). Kolkata, India.
24. Jonathon T. (2019), Lean Manufacturing Explained. Available: https://www.reliableplant.com/Read/30267/lean-manufacturing-techniques (accessed on 1 December 2020). Kolkata, India.
25. Makkar, S., Devi, N., & Solanki, V., (2019). Applications of machine learning techniques in Supply chain optimization. *International Conference on Intelligent Computing and Communication Technologies (ICICCT 2019).* Springer, Singapore.
26. NIBusiness (2019). *Artificial Intelligence in Business*, URL: https://www.nibusinessinfo.co.uk/content/risks-and-limitations-artificial-intelligence-business (accessed on 1 December 2020). Kolkata, India.

CHAPTER 3

F-COMMERCE AS A TOOL OF ARTIFICIAL INTELLIGENCE: A MODERN APPROACH FOR PROMOTING STARTUPS

PRIYANKA JINGAR[1] and SACHIN GUPTA[2]

[1]*Research Scholar-Doctoral Studies, Department of Business Administration, Mohanlal Sukhadia University, Udaipur, Rajasthan, India*

[2]*Assistant Professor, Department of Business Administration, Mohanlal Sukhadia University, Udaipur, Rajasthan, India, E-mail: sachinguptabusadm@gmail.com*

ABSTRACT

Artificial intelligence (AI) is expanding and sprouting faster than ever. It is a swiftly mounting technology, made possible by the Internet. AI has a significant impact on digital commerce. Facebook commerce (F-commerce) refers to digital commerce that uses the social media platform to capture the masses. Facebook represents a frequently sprouting marketing stage and communication pathway. Worldwide, there are over 2.38 billion monthly active users, and India has the highest number of Facebook users in the world. Facebook has become a strategic element of marketing and business promotion activities. Such platforms are a boon chiefly for startups having controlled budgets. Facebook is all about people, social selling, and relationship building. It's easy to access, low-cost marketing platform, connecting with existing and potential customers, targeted advertising, raise brand awareness, and positive word of mouth has proven to be the key to success for startups. The current chapter will throw the lights on the features of Facebook in the form of AI. The purpose of this chapter is to highlight the role of AI to create a new marketplace for all social media users.

3.1 INTRODUCTION

In a globe filled with multi-platform companies, it's tough to keep up with the competition if we don't reach where our customers hang out largely, and Facebook is an online platform where people spend the majority of their time. Facebook users spend an average of 40+ minutes per day on the popular social network and a good portion of that time goes towards Facebook shopping. This is the foremost reason why a business on Facebook is such an amazing thought. The most attractive thing about the Facebook marketplace is that we do not need to have any pre-existing physical store and can run the whole operation via Facebook. But the perks don't end here, and Facebook marketplace offers a mixed bag of services, which includes:

- The infinite number of products;
- Organize products into collections and categories, for example, home and garden, clothing and accessories, electronics, family, classifieds, jobs, hobbies, entertainment, films, property, etc.;
- Communicate with customers directly via page;
- See statistics on sales, visits, and more;
- Get the products appearing in Facebook Marketplace, giving access to a much larger base of potential customers.

3.2 REASON TO CHOOSE F-COMMERCE

Facebook commerce (F-commerce) refers to carry out online business activities on a Facebook page or Facebook application. Facebook marketing is an exercise of encouraging a brand and sustaining its presence on Facebook. Promotion on Facebook is both organic and painful. It is the most important platform for startups to promote their product and services, create brand awareness, reach to masses, identify their target customers, etc. Startups that concentrate less on direct selling and instead use artistic ways to unify into the experience users like most about Facebook will make a larger crash with their promotion. Startups must consider the below-mentioned points for choosing Facebook for promotions:

- As of June 2019, Facebook has a monthly active user base of 2.41 billion users worldwide.
- 1.59 billion people log into Facebook daily and are considered active users in June 2019.
- 42% of marketers account that Facebook is significant to their business.

- As of May 2013, 16 million Local business pages have been created on Facebook.
- Five new profiles are created every second.
- Once Facebook users login, they spent an average of 20 minutes per visit [13].

3.3 OBJECTIVES OF THE CHAPTER

1. To know about various new features of Facebook in the form of artificial intelligence (AI).
2. To identify the usage pattern of Facebook by entrepreneurs.
3. To assess the advantages of using Facebook by entrepreneurs.
4. To find out the effect of using Facebook on startups.
5. To figure out the contribution of F-commerce in the promotion of startups.

3.4 FACEBOOK: CONNECTING PEOPLE

Facebook mission: "Give people the power to build community and bring the world closer together. That reflects that we can't do these ourselves, but only by empowering people to build communities and bring people together" [18].

Facebook was founded by Mark Zukerberg and fellow Harvard college students in 2004. Facebook service can be accessed with any device having internet connectivity, such as mobile phones, tablets, personal computers. Facebook is a well-liked free social networking platform that permits registered users to make profiles, upload photos and videos, send messages via Messenger and keep in touch with friends, family, and colleagues, create a marketplace, etc. It is widely available in 37 different languages and includes the following public feature such as:

- Marketplace-Facebook permits its members to post classified ads, read, and respond to those advertisements.
- Groups-It provides a platform for its members to find, connect, and interact with each other that are having a common interest.
- Events-allows members to broadcast an event, call guests, and track who plans to attend.
- Pages-allows members to generate and promote a public page built around a definite topic.
- Presence technology-Facebook permits online members to chat.

F-commerce is buying and selling goods and services on the Facebook platform. It is one of the world's most important companies. It is considered as one of the giant technology companies, along with Google, Apple, and Amazon. Facebook is the most extensively used social media platform in the globe, with more than 700 million worldwide users. Facebook was designed to be very user-friendly, catering to virtually any person with even the most fundamental of computer skills. For those with a piece of advanced programming knowledge, applications, and supplementary features are presented for a more customized look, feel, and functionality.

The initial hypothesis behind the site's development was to unite a limited group of individuals with resembling interests. While the website was formerly created to connect individuals, but nowadays, all business institution, non-profit organizations, groups, communities, and public figures are using Facebook for promoting brand awareness and customer satisfaction with the help of various features such as marketplace, trending photos, and videos, enhancing network connections, creating groups, etc., (Figure 3.1).

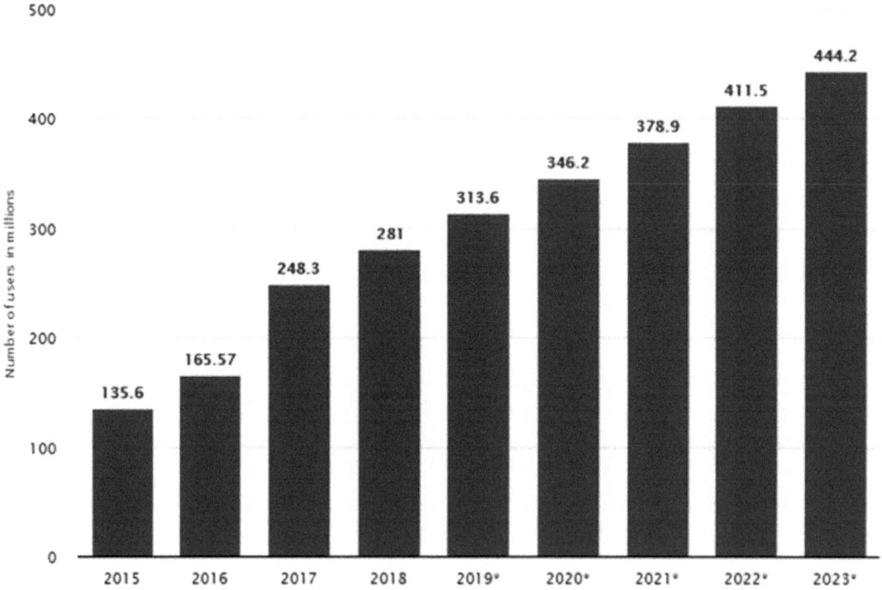

FIGURE 3.1 Number of Facebook users in India from 2015 to 2023 (in millions).

Source: Reprinted Hutchinson [10]. © 2019 Industry Dive.

The number of Facebook users is rising every day, every month, every year. There is an expedited growth of users from 2015 (135.6 million) to 2019

(313.6 million); however, Facebook expects a whopping growth in several users till 2023, which shall be 444.2 million users.

3.4.1 ACTIVE ADVERTISER ON FACEBOOK

In the first quarter of 2019, Facebook declared that over 7 million active advertisers were using the social networking platform to promote their products and services to create brand awareness and customer satisfaction [3].

3.5 ARTIFICIAL INTELLIGENCE (AI)

AI is something that eases our day-to-day life. The spectrum of AI covers innovation and technology. It is an imitation of human intelligence processes by machines and technology, especially via computer systems. It is the aptitude of a digital computer to execute all the tasks usually allied with intelligent beings. The best example of AI is Siri, Google Assistant, Facebook Messenger, etc.

3.5.1 ROLE OF ARTIFICIAL INTELLIGENCE (AI) IN FACEBOOK COMMERCE (F-COMMERCE)

"Sky is the limit" this quote has flawlessly applied in the grassland of innovation and technology. Technology has gained an altitude where a human being can accomplish implausible targets. Every person wishes to get things done very swiftly and efficiently. The prospect of learning is allied with the development of AI.

It is an era of social media. F-commerce is a subset of social commerce. Whether it is about reaching out to new clients or nurturing existing business relationships, Facebook has become a vital ingredient of every business. Effective Facebook management helps enhance a brand's strength and building customer relationships for startups. It is a proficient way of engaging millions of social media users. AI helps analyze baggy data to make out trending subjects, hashtags, and patterns to recognize user activities. A variety of companies and startups that offer online marketing services are discovering fresh ways to use the F-commerce platform with AI. Companies and startups have started using AI to discover new demographics to arrive at a targeted audience.

Several social networking giants such as Facebook, LinkedIn, Twitter, Alibaba, Instagram, WhatsApp, and Amazon have started using AI techniques to augment profound learning ability. The use of AI has facilitated to compel innovation in numerous associated pitches. AI is hurriedly transforming the approach social big data is used for insights. AI in F-commerce is gaining eminence. Messenger apps, such as Facebook Messenger and WhatsApp, are not just only for private conversations any longer but have become social media marketing platforms in themselves, providing access to both commerce and service applications.

3.5.2 FEATURE OF FACEBOOK IN THE FORM OF ARTIFICIAL INTELLIGENCE (AI)

1. **Provide a Personal Touch with Chatbots:** The rise of Chabot is one of the supreme sensations of AI. AI's integration with Facebook will help brands to develop some cool chatbots and then place them on their account to assist users in interacting with brands in a more modified way, for example, Facebook Messenger, Siri, Google Assistant, etc. These AI-powered chatbots act like a 24×7 customer support assistant to identify customer's problems and facilitate customers just like a salesman. Glimpses of chatbots are shown in Figure 3.2.
2. **AI-Based Image Recognition:** In the era of social media, images are valuable tools to engage people and can help identify consumer behavior patterns. AI-empowered software recognition tools can help congregate actionable insights to recognize the shift in user patterns through millions of images posted on social media. This feature will help startups to identify their targeted audience.

   The facial recognition feature of Facebook is entirely based on machine learning (ML). This feature helps business startups a lot to post their products with labeling and product specifications. It is again a smart way for Facebook to recognize the people, tag them, notify them with the latest updates, etc. (Figure 3.3).
3. **Facebook Marketplace:** Introduced in 2016 as a platform for users to purchase and sell objects within their local society without any charges. In the U.S., one out of three users on Facebook use the marketplace for buying and selling products and services [17]. Whether a consumer wants to buy a product or a company wants to sell, AI is behind the curtain and making user experience effortless.

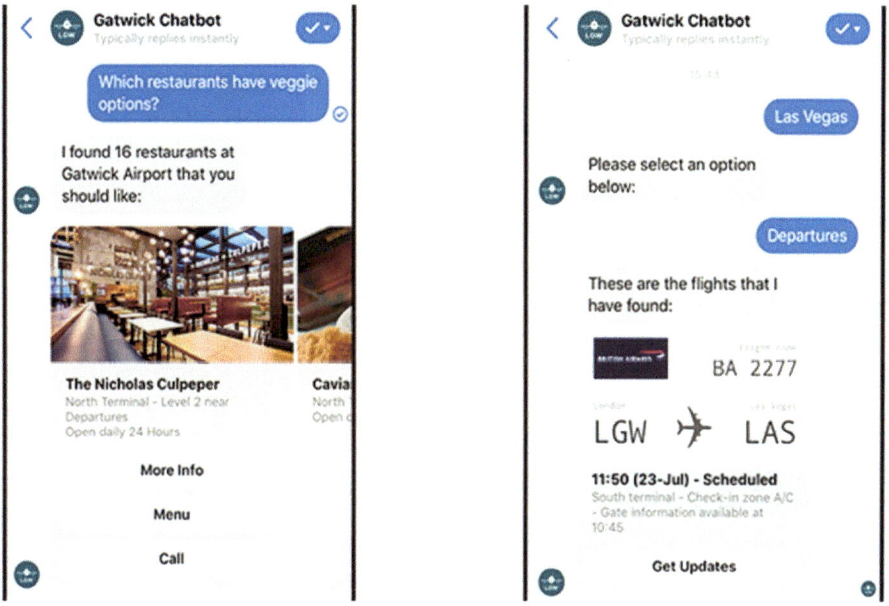

FIGURE 3.2 Glimpses of chatbots.
Source: Reprinted from Wearables [19].

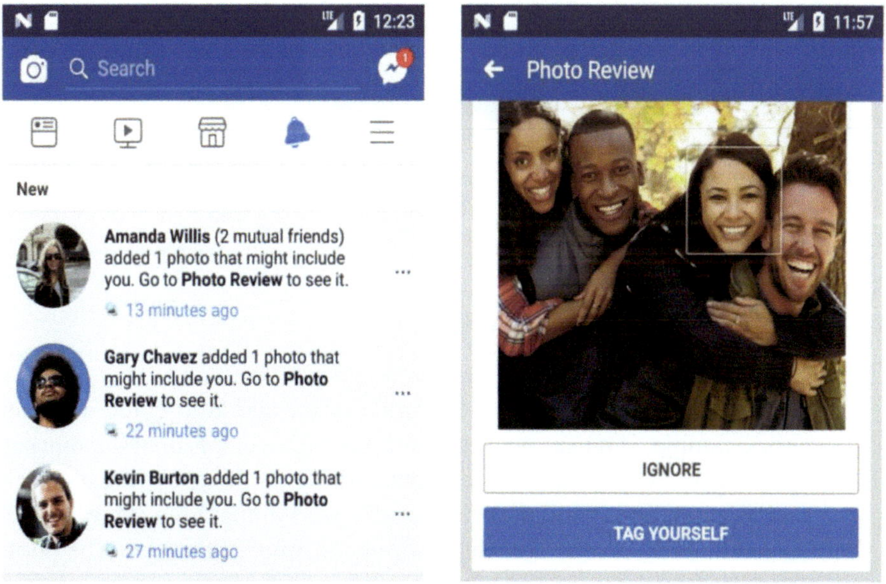

FIGURE 3.3 Facial recognition feature of Facebook.
Source: Reprinted from Kahn [12]. © 2018 Stuff Limited.

Facebook Marketplace has launched Pan India; it became a threat for Olx, Quikr, etc., which offers second-hand products to the consumer. Especially since Facebook is present in 1,000 cities in India and functions across 14 industries, including mobile phones, fashion, household goods, cars, real estate, jobs, services, and education, etc., [1]. AI helps the buyer to locate items. AI is intended to improve its user interface through personalization, faster transactions, and improved safety measures (Gallagher). Perks for startups are extended reach to massive customers, personalized customer experience, and no listing fee (Figure 3.4).

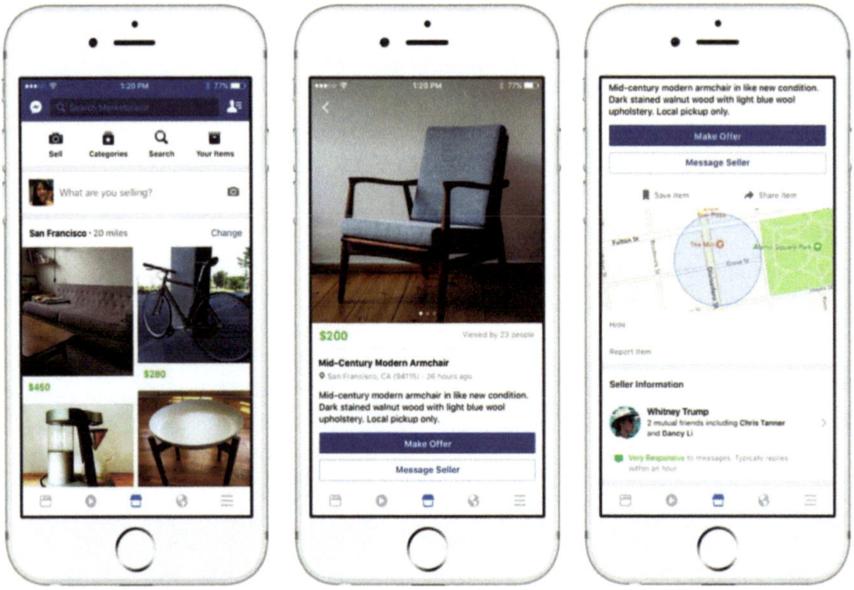

FIGURE 3.4 Facebook marketplace.

Source: Reprinted from Constine [4]. © Verizon Media.

AI also helps to suggest the price range and auto-categorization. For example, if a user wishes to sell his office chair, then the marketplace uses AI to assist the customer in fast-selling by suggesting a price based on a similar chair recently sold in that area. Marketplace mechanically categorizes the chair as "furniture" based on the photo and description.

4. **Expressive Interactions:** In the first year of launch, Facebook emoji were used over 300 billion times, which is 800 million a day [15].

Facebook allows users to react to the post with six different emotions, that is Love, Wow, Haha, Sad, Like, and Angry. Since 2017, Facebook gives more magnitude to the posts having more emotive reactions than just a simple 'like.' With the help of AI, Facebook ranks up those posts which have additional interactions and will display case that post in the newsfeed of further people as per their preceding likings.

Facebook emoticon facilitates startup businesses to recognize how the post makes the user feel. This information can assist startups in comprehending their customers better. It instigates startups to get the sales-focused approach. Psychologist Daniel Kahneman says that people have a propensity to be more emotional than logical (Figure 3.5).

FIGURE 3.5 Facebook emoji.

Source: https://www.freepik.com/free-photos-vectors/emoticon.

It is a smart way to know the customer's reaction to the brand. It is a largely influential tool for startups to collect data regarding consumer experience. Any startup needs to get a review of the brand, product, and services. The AI algorithm furnishes startup businesses a chance to listen and respond to user reactions. The consumer feels delightful, and the business gets constructive feedback. Startups can keep an eye on who is indulging with the brand and how they're indulging. Such data can be used to optimistically profile customer experience.

5. **Ranking the Content:** Many social platforms like Facebook, Instagram, YouTube, and Twitter gather user's data by using AI. It receives a massive amount of unstructured statistics in the forms of likes, comments, favorites, views, and shares to generate a realistic statistics-showing result, retweets, screening the mainly probably likable ones at the pinnacle. AI features in Facebook help enhanced interaction with users (Figure 3.6).

FIGURE 3.6 Facebook enhanced interaction with users by using artificial intelligence features.
Source: Reprinted Patil [14]. © 2019 ESDS Software Solution Pvt. Ltd.

The following things may affect how video is ranked in the news feed (Table 3.1):

i. **Originality:** Facebook will more powerfully bound allocation and monetization for copied or repurposed matter.

ii. **Video and Viewing Duration:** Facebook will build up the ranking indicator of videos that keep people occupied for a comprehensive period.

iii. **Loyalty and Intent:** Facebook will reinforce the ranking indicator of video that people come across for and revisit over and over [16]. Facebook prioritizes videos that users look for via search option or by visiting a company page.

TABLE 3.1 Formula of Ranking on Facebook News Feed (Visibility = R + I + P + B + U + T + T)

Recency	How the latest post was published.
Interaction	Interaction level between customer and company. It is based on chat history, the number of comments, share, tagging, etc.
Performance	Performance of the post amongst other users.
Buddies Activity	The reaction of a friend or customer on the company's post.
Uniqueness	The fresh content on Facebook is given superior priority. Content should be exclusive.
Trending	The topic of the post is trending or not.
Type	Type of the post such as video, photo, link, status, etc.

6. **Improve Recommendations for Customers:** Not every customer wants the same thing, so AI provides a customer-centric search. For example, Amazon makes recommendations to users depending on their activities on the site and any past purchases. Netflix makes TV and movie recommendations based on a user's interaction. On Facebook, people use recommendations to ask friends in groups: where to go, where to shop, where to hang out, where to eat, etc. Facebook reviews are becoming recommendations for startups as well as customers. Recommendations are not only the outlook given by users openly on the business's Facebook page but also when people respond to their questions when they look for a recommendation on Facebook (Figure 3.7).

Reviews and recommendations are tremendously essential for any company's status. It is very appealing for startups to know that 1 in 3 Facebook users are using the stage to find reviews and recommendations. For example, Maps Recommendations are discoverable on Facebook when users are searching for or talking about the business. The paramount thing is that Facebook shows the reviews of the nearest user to sanction the purchasing decisions and fetch greater faith [2].

When a user drops a recommendation on the Facebook page directly, then the user has the choice to choose a tag that describes their recommendation better. These tags are in the form of the hashtag "#." A startup business can grow its business with Facebook recommendations.

FIGURE 3.7 Facebook recommendations for customers.

Source: Facebook business (https://www.Facebook.com/business/recommendations).

7. **Building Relationships:** The purpose of Facebook is to provide everyone with a platform where they can connect with people. A marketer can make a group of target customers based on their common interest to interact, creating brand awareness.

3.6 RESEARCH METHODOLOGY

As the research work is conceptual, the researcher has applied exploratory research design by using various secondary data availed from the secondary data sources. A research report, journal, and newspaper article from renowned writers have been reviewed. Based on the secondary data and review, the researcher has reported on a mixture of emerging trends in F-commerce with the help of AI and how AI is bringing the world closer together on a single social media platform.

3.7 SIGNIFICANCE OF THE CHAPTER

In the present era of cut-throat competition, organizations are constantly seeking to increase their market share along with maintaining the existing with the help of different types of media such as print media (newspapers, magazines, etc.), electronic media (radio, television, etc.), outdoor media, cinema, mobiles, and the Internet. To attain the organizational goals, marketers are applying a proper mix of all media vehicles. Nowadays, the Internet is becoming a very popular, effective, and low priced model of advertising and promoting products and services. In this course of action, F-commerce is emerging as a very powerful tool for marketing products and services and convincing the customers over the Internet. Today, most organizations are seeking to utilize this innovative approach of reaching the existing consumers and making the prospects aware of their offerings all over the world. This approach is called F-commerce.

Facebook has started as a clean social media-networking platform to facilitate people to hook up with friends and family, relatives, but now it has emerged as an effectual means for advertising brands and promoting businesses. It is for this reason that marketers are constantly looking to classify means to connect with budding customers. With this approach, Facebook begins to hook up people with people and people with ideas. AI has gained tremendous consideration in recent years. Innovation, made achievable with the help of the Internet. AI is a technology that is already impacting the user's interaction with each other. In a few years, its collision is likely to only persist in growing up. AI has the power to immensely transform the approach that people interact with, not only with the digital era but also with one another. Facebook is a platform that uses AI to reach the masses. *Facebook* is one of the most efficient ways *to advertise* online, to create a marketplace, promotion, and newsfeed, etc. A major benefit of Facebook advertising is its ability to reach the exact audience. Facebook is now an imperative ingredient of every business's marketing strategy, to produce leads and renovate them into reliable consumers. Facebook is the most targeted form of advertising. The seller can advertise to people by age, interests, behavior, location, etc. This study will help to identify the scope of F-commerce in India and how it is changing the orthodox trend of advertising and promotion.

3.8 ADVANTAGE OF FACEBOOK COMMERCE (F-COMMERCE) TO STARTUPS

1. **Massive Exposure on Global Scale:** Facebook is undeniably the player of social media. It is towering over the rest of the social media with a tremendous increase in the number of active users. Facebook proposes numerous platforms for promotion in the shape of pages,

groups, and ads. Facebook page for an individual or a startup is the trendiest way to signify an individual or a business. A group page can be released by a startup to uphold activities. Users can join this group and reviews it in the form of recommendations. It gives massive exposure to any startup to stand out in the crowd. Facebook has a considerable quantity of demographic information of all of its users, and the ads are planned to target users with particularly precise demographics preferred by the advertiser. For example, the "Dove Real Beauty Sketches" campaign was started by Dove Company in 2013. Within the first 10 days, the movement got shared over 6,30,000 times on Facebook. This is the massive exposure that relies on companies [11].

2. **Lower the Marketing Expenses:** Promotion or advertisement on Facebook is relatively inexpensive as compared to the traditional method of advertisements such as Print media, TV, Radio, Hoardings, and board, etc. With the help of photos and videos, it becomes tempting for a user to watch. Creating a website for small businesses require money. On the other hand, a Facebook business page can be created without a single penny, and any content can be uploaded by a startup on this business page. Even a small startup can start their business via a Facebook page and display the relevant information regarding its products and services free of cost.
3. **Capacity to Target Potential Audience:** Facebook is a social community of people belongs to a different age, income, interest, gender, etc. With the help of AI, a startup can reach its target audience for whom he is producing goods and services. If a startup wishes to post an ad for youthful business owners, the startup can select demographics of age below 35, resides in metro cities, middle-income group, etc. In addition, to publicize female fashion, a startup can target women with the age group between 20 to 50, with a keen interest in fashion and trendy apparel, belongs to the local area, etc. Such specific and refined filters are very beneficial for any business to identify its target audience.
4. **Brand Awareness:** Facebook is one of the largest social media platforms in the world. A business presence on Facebook will help the startup to gain exposure to potential customers. Besides being a platform to construct a customer base and advertise products, a Facebook business page can do marvel for helping startups to construct brand loyalty. Facebook promotes effortless and speedy interaction between customers and the company. It is also easy to react to

consumers, which encourages brand loyalty. A swift customer interface means it indicates efficient customer hold by giving a reply. Facebook also builds commitment. Commitment in the form of likes, comments, reviews leads to a strong bond. The stronger the relationship with the customer, the more likely it will persuade sales.

5. **Increase in Web Traffic:** A startup can post a photo with the link. By clicking on the link, a user can directly redirect to the website. It is done by a thumbnail image. It creates an online visit of the customer, which leads to growth in the selling of products. Jabong started a campaign on Facebook in 2017 and reported a significant boost in sales by reaching out to the right people with the right product.
6. **Build Long-Term Relationships:** Facebook page is a truly grand way to integrate information and also connect with the customer. Unlike faster platforms like Twitter, or image-oriented ones like Instagram, Facebook gives the room to build a long-term, interactive platform [9].

3.9 CONCLUSION

Worldwide, the current active population on the Internet is crossing 4.3 billion, out of which 3.4 billion is the number for social media users [14]. Every startup wants to connect with social media to make their business reach new tallness and amplify sales. Earlier Facebook was simply a social networking site on which friends and communities stay in touch, but nowadays, many startups and brands are using AI to make their presence on Facebook. Facebook was currently approaching 2.41 billion user bases in 2019. No other social media network is nearby Facebook. "Facebook is now big enough to support further commercialization." Facebook is becoming a new online advertising giant next to Google. The introduction of AI boosted the performance of F-commerce.

AI is everywhere nowadays. Social media is one such domain that is being affected by AI development. With upwards of 3 billion people logging into social media sites regularly, the procedure of optimizing social media marketing has received a massive boost through the application of AI. One of the rewards of AI is that it acts as a high-performance digital brain that provides direction to the marketer, startups and helps them generate a more successful strategy for the practice of customer targeting, creating brand awareness, and customer satisfaction. The long-established notion of commerce was the right product must be available at the right place and at the right time is now replaced with the elastic concept that is product or services should be available anywhere and anytime. This vibrant theory has instigated the need for F-commerce in India. With the help of AI, F-commerce has revolutionized the experience of customers and marketers, and startups. It leads to build brand awareness, healthy customer relationship, and pump the sales. Therefore, the contribution of AI in F-commerce for promoting startups is tremendous. Hence, AI plays a critical role in maximizing engagement by studying the behavioral pattern.

KEYWORDS

- **artificial intelligence**
- **digital commerce**
- **Facebook commerce**
- **promotion, startup**
- **social media**
- **web traffic**

REFERENCES

Anupam, S., (2018). *Facebook Intensifies Facebook Marketplace: No Deadline For Pan-India Launch.* Retrieved from Inc 42: https://inc42.com/buzz/Facebook-marketplace-pan-india/ (accessed on 1 December 2020).

Bassig, M. (n.d.). *Establish your Reputation with Recommendations.* Retrieved from Facebook Business: https://www.Facebook.com/business/recommendations (accessed on 1 December 2020).

Bojkov, N., (2018). *New Facebook Recommendations and Reviews System.* Retrieved from embed social: https://embedsocial.com/blog/Facebook-recommendations/ (accessed on 1 December 2020).

Clement, J., (2019). *Number of Active Advertisers on Facebook from 1st Quarter 2016 to 1st Quarter 2019 (in millions).* Retrieved from Statista: https://www.statista.com/statistics/778191/active-Facebook-advertisers/ (accessed on 1 December 2020).

Constine, J., (2016). *Facebook Launches Marketplace, Friendlier Craigslist.* Retrieved from: https://techcrunch.com/2016/10/03/Facebook-marketplace-2/ (accessed on 1 December 2020).

Facebook emoticons. Retrieved January 03, 2021, from Freepik: https://www.freepik.com/premium-vector/facebook-emoticons-collection-with-different-faces_2235879.htm.

Facebook's New AI Features to Help 'Marketplace' Shoppers, Sellers, (2018). *Tech News 18.*

Gallagher, K., (n.d.). *Facebook Marketplace is Getting an AI Facelift.* Retrieved from Business Insider: https://www.businessinsider.com/Facebook-marketplace-artificial-intelligence-2018-10?IR=T (accessed on 1 December 2020).

Hive, T., (2017). *Fourteen Reasons to Use Facebook for Business.* Retrieved from TriveHive: https://thrivehive.com/benefits-of-Facebook-for-business/ (accessed on 1 December 2020).

Hutchinson, A., (2019). *Facebook Reaches 2.38 Billion Users, Beats Revenue Estimates in Latest Update.* Retrieved from SocialMediaToday: https://www.socialmediatoday.com/news/Facebook-reaches-238-billion-users-beats-revenue-estimates-in-latest-upda/553403/ (accessed on 1 December 2020).

Jhajharia, R., (2018). *Six Unbeatable Benefits of Facebook Marketing Your Business Should Consider.* Retrieved from Digital Vidya: https://www.digitalvidya.com/blog/benefits-of-Facebook-marketing/ (accessed on 1 December 2020).

Kahn, J., (2017). *Facebook Introduces New Facial Recognition and Privacy Features.* Retrieved from Stuff: https://www.stuff.co.nz/technology/social-network-ing/100026806/Facebook-introduces-new-facial-recognition-and-privacy-features (accessed on 1 December 2020).

Noyes, D., (2019). *The Top 20 Valuable Facebook Statistics.* Retrieved from Zephoria Digital Marketing: https://zephoria.com/top-15-valuable-Facebook-statistics/ (accessed on 1 December 2020).

Patil, D., (2019). *Role of Artificial Intelligence in Driving Social Media.* Retrieved from ends: https://www.esds.co.in/blog/role-of-artificial-intelligence-in-driving-social-media/#sthash.13iUxm6h.dpbs (accessed on 1 December 2020).

Russell, J., (2017). *Facebook Reactions: What They Are and How They Impact the Feed.* Retrieved from HootSuite: https://blog.hootsuite.com/how-Facebook-reactions-impact-the-feed/ (accessed on 1 December 2020).

Southern, M., (2019). Facebook changes how videos rank in the news feed. *Search Engine Journal.*

Wearables (2019). eMarketer, January 3, 2019.

Zheng, L., & Li, R., (2018). *Under the Hood: Facebook Marketplace Powered by Artificial Intelligence.* Retrieved from Facebook Engineering: https://engineering.fb.com/ml-applications/under-the-hood-Facebook-marketplace-powered-by-artificial-intelligence/ (accessed on 1 December 2020).

Zuckerberg, M., (2017). *Bringing the World Closer Together.* Retrieved from Facebook: https://www.Facebook.com/notes/mark-zuckerberg/bringing-the-world-closer- (accessed on 1 December 2020).

CHAPTER 4

SECTORIAL IMPACTS OF ARTIFICIAL INTELLIGENCE

OM PRAKASH GUSAI[1] and ANKUR RANI[2]

[1]*Assistant Professor, Department of Commerce, Motilal Nehru College, University of Delhi, New Delhi–110021, India,*
E-mail: opgusai@mln.du.ac.in

[2]*Research Scholar, University of Delhi, New Delhi–110021, India*

ABSTRACT

Artificial intelligence (AI) is proving to be of great help to humankind in many fields. It can complement human efforts in these fields by analyzing the behavior of transactions, anomaly detections, and tracing the major solutions. AI performs, or we can say it is that type of machinery that mimics the "cognitive human mind" like giving solutions to the problems and different learning aspects. This chapter mainly focuses on the impacts of AI in various major sectors of the economy. AI becomes a hot issue because of its various applications that replace the position of a human being to some extent. The present chapter highlights the use of AI in accounting, auditing, marketing, human resource management, and major in the banking sector. The existing patterns of fraud detection lay down on the predefined criteria of specific rules that make detection techniques very difficult, so new AI techniques and models have been adopted to find suspicious transactions and fraudulent acts in various fields.

4.1 INTRODUCTION

In today's environment where everyone is involved in the digital world somehow, we can say the use of artificial intelligence (AI) forms a place in the human mind. AI works like an umbrella that covers many technologies from the root level to a sophisticated level. According to Auden, a poem based on human vs. robots is "those who will not reason Perish in the act; those who will not act Perish for that reason" it refers to the fittest for survival. Those who can tackle the problem will survive for a long. It works like a platform where machine learning (ML) and its discoveries find super importance. Right now, AI is a hot issue which has taken place in humans' life [1].

4.2 MEANING: ARTIFICIAL INTELLIGENCE (AI)

"Nils J. Nilsson separates AI into its components, which are as Artificial (machine, as opposed to human) + intelligence. To gauge the intelligence of an entity, Nilsson falls back on the Turing Test. Moreover, he believes that with the increased complexity of the machine comes increased intelligence [2]."

The word "AI" was coined by John McCarthy. It is an experimental part of computer science where it expels the diverse use of machines. There are different activities that computers can perform using AI techniques can be recognition of speech, attainments in auditing, structural planning, and judicious tasks.

4.3 TECHNIQUES OF AI

The formation of the AI model includes different verities of supervised and unsupervised techniques for finding out correlated data:

1. **Supervised Algorithm:** This type of algorithm is used to learn some complex patterns. In this, the output is already known before setting up the input in the processing system. The raw data is processed with the help of a particular supervised algorithm to obtain the desired output.

   The various types of algorithm used under supervised models are as follows:

   i. **Logistics Regression:** It is part of the statistics field. This method is used to transform the output prediction using logical functions and identify the cause and effect relationships among the ocean of data. Its efficiency and less computational tasks make their use widespread [7].

ii. **Learning Decision Tree:** This method is more suitable than others. For modifying customer behavior, a specific set of rules is used. It is also known as a classification tree in which the input characteristics are labeled with each node.
iii. **Random Forest:** As like in the forest, those trees which grow in very deep, that easily pertain to the irregular patterns in which grow in various directions. This method helps to average the deep multiple decision trees and work on the same data set but different parts. The main aim of random forests is variance decreasing to some extent. It is a boosting technique because it ensembles various classifiers that are weak into a combination of strong classifiers [3].
iv. **Neural Network:** This method detects fraud in real-time. As it works like a human brain made of multiple nodes and link with others which perform simple tasks and their output is known as activation/nodal value passed to other neurons able to find normal behavior patterns.

2. **Unsupervised Algorithm:** This algorithm takes place when the particular output is unknown for the system because the data used here is a concrete type. It is used for a limited number of applications and detecting anomalies.
   i. **K Means Clustering:** It is an unsupervised method that helps to design a cluster in which the n observations are being separated in k types of clusters that is the combinations of neighbor points that serve as a prototype (Figure 4.1).

      In the above diagram, the n number of different observations is separated into k types of clusters.
   ii. **Local Outlier Factor:** The local density can be measured by k nearest points of data sets, which are isolated with the particular object on which anomalies are dependent. These anomaly scores are known as local outlier factors. Outliers are transactions that contain fraud factors. It helps to identify the low-density samples as compared to the neighbor's points that already have local points.

4.4 AI: A LAYPERSONS' VIEW

In the eyes of a layperson, understanding of AI is limited to its uses like lower skills of technology, computer, mobile-related aspects through which comparison between high skilled technology users and others. Normally people don't know what AI means? Mostly higher-skilled people understand in a better way about AI uses and functions as compared to lower-skilled people. It is as like our old generation people generally don't believe in machines works but today's generation say even a common man believes that their work can be done by a ROBOT or any ML process [3].

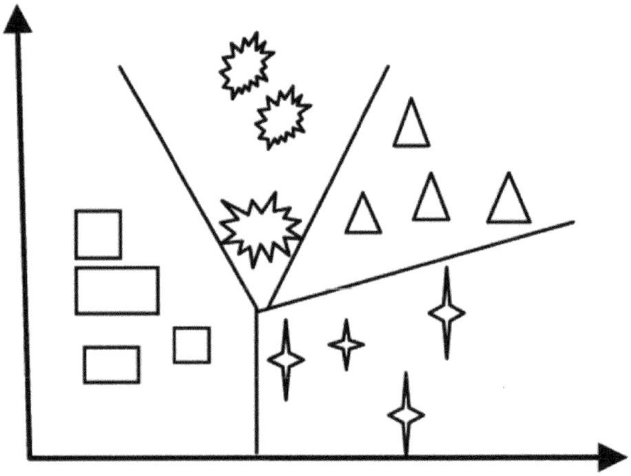

FIGURE 4.1 K means clustering.

4.5 FACTORS INITIATING AI

As for upcoming years, AI will bring a drastic change in human life. There are mainly three factors that initiate machine-learning importance in our lives (Figure 4.2):

1. **Computing Structures:** These in the old days were based on central processing units. These processors and CPU were created, and their designing patterns were not based on models of ML. The fastest processors were also not able to find the proper ML models. The forgiving major advancement in applications of AI, there must be some improvement in CPU processor, which arises a new term that is known as 'GPU' (graphic processing unit). It was considered as a constituent of workstations is now a people cloud after sought processing. GPU contains various typical points that speed up the artificial learning process. In the public cloud, various devices of customers, and the machines which are useful in daily works, GPU

is the main part and is key to ML. There are new inventions that occurred after GPU that is "Field Programmable Gate Array." These types of processors are designed for niche tasks related to computing and customizable for ML models.

2. **Neural Network Expansions:** These are the critical factors that questioned the rise of AI. "Artificial neural networks (ANN)" takes the place of old "ML models." From deep learning to the better vision in computing tasks like "single shots outbox detector" for making an image. This is used for increasing the quality of the picture in high definition, which is in 'low resolutions.' The other sectors of the economy also get some ideas from these types of neural networks (NN).
3. **The Historical Data:** This was unlocked by some companies and government in 'magnetic disk,' and recorders can access a huge variety of expansion after training in ML models. For finding the solutions to a big problem, there is a need for large datasets points. With the use of AI, the unstructured data can be retrieved inexpensively. It is a factor that signifies research in AI.

As these are the crucial factors that initiate the need for AI in the fastest manner for originating innovations, various companies like "IBM," "Facebook," "Google" are investing huge money and pays for a wide area in research related to different aspects of AI.

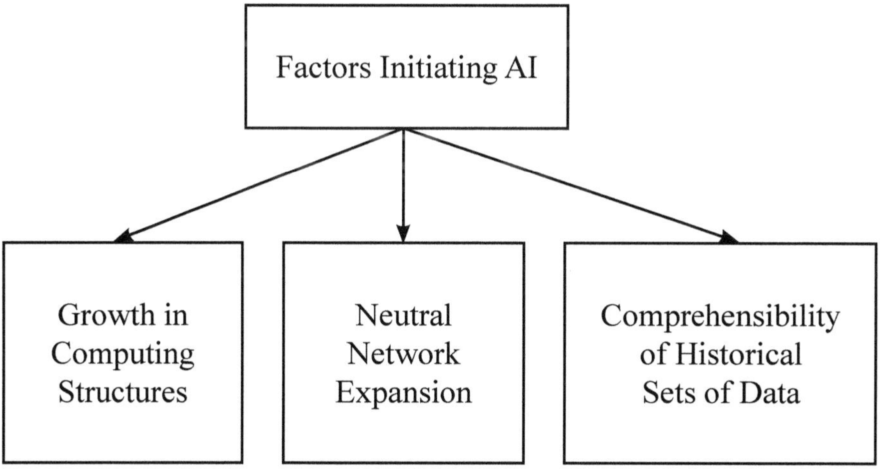

FIGURE 4.2 Various factors which arise in AI.

4.6 DIFFERENT SECTORS WHERE AI PERFORMS

4.6.1 ACCOUNTING AND AUDITING

AI gives a platform for accounting firms to explore and grasp the latest technology-driven aspects. Some of the useful accounting tasks AI performs are listed below:

1. It helps companies by obtaining, consolidating, and merging data from different and numerous sources. This saves time and helps them to plan their goals effectively.
2. AI helps in searching and accessing digital files easily. Now, this increases the precisions of the auditing done as a firm's financial transactions can be looked upon easily.
3. It can review receipts, expenses and identify if there is any breach of accounting policies and procedures.
4. It helps resolve queries of the consumers and keep track of their account balance, pending bills, etc.

Interrogating the Auditing functions through AI can decrease the mammoth size of work from both client and company sides. In auditing, the analysis of ledger or financial reports or any manual efforts that can be minimized by AI and auditors can provide a better context than before. An auditor can gather useful data with ML for identifying the repercussions and praiseworthy aspects related to accounting terms and risk solving factors.

4.6.2 MARKETING

As we all know that in a particular organization, the place of marketing tasks is on the upper part. The main aim of any organization is to satisfy its consumers and fulfill their desires. With the help of AI, a marketer obtains those approaches which are consumer-oriented and understands the main fact about where the consumer is lacking to make decisions or giving a fruitful way to work in a 'decision environment.'

In marketing, "making decisions" in the right manner is a crucial task. It requires a lot of intuition and judging power in a marketer. As the complexity comes in day-to-day transactions, the person's decision-making capability seems to be difficult. The machine we can say the engine of marketing decisions can be neutral the noise. The AI system provides all the forecast data. Classifying the present trends, providing information daily. AI explores the 'decision support system' via 'collaborative decisions.' The ML techniques and their applications are drawn for a "decision support system" and usually aid all uncertainty relating to decision-making tasks [4–7].

4.6.2.1 USING EXPERT ADVICE IN MARKETING

For making the use of AI in marketing, we generally use some "expert system" that is called software's for problem-solving in experts' language can be like providing useful judgments and giving advice for any specifics cases which normally a marketer is not able to handle. The main examples of "expert system" software for the marketing area can be "markex" and "brand frame."

The former provides consultancy to marketers and useful in every stage of the decision-making process, the production-related tasks, for consulting to make decisions in pricing model or any other task of marketing decisions. The latter helps to take decisions and to compare brands, classifying the attributes of brands, the main aim, and the budget capacity to promote a brand for a brand manager. As the input is feed by the manager in this software, this reacts according to the input and recommends whether the company is fitted for promotions campaigns or should there be a need for any improvement. The use of AI in marketing provides a platform to tackle problems in an easy way for a marketer.

4.6.2.2 DIGITALIZE MARKETING VIA AI

In today's digital era, where most tasks of marketing fulfill by assistive techniques, we can measure the main effect in "social media." Every business enterprise performs tasks using the social media platform and forms a type of "asset" for marketing development. Here, we are going to explain some strategies which use AI as a digital tool.

4.6.2.3 DIFFERENT TYPES OF MARKETING STRATEGIES [8]

1. **Continuous Marketing:** Reminding customers continuously is a big task for companies. For making goodwill and keeping its value

in the market, companies use continuous marketing strategies for making their customer in a long relationship and need to balance all the promotional campaigns offline for getting fruitful results as customers are the only assets of the companies, which increases its value.

2. **Personalized Marketing:** Marketing through televisions and newspaper media becomes older, and now companies use personalized marketing to know the customer better. The use of software which easily tells about the priorities of customer towards a particular product like Myntra, Amazon, flip kart all companies easily find out what customer wants and how personalized marketing enhances information relevant.
3. **Mobile Marketing:** 'Mobile' which is almost becoming a necessary item in daily life, mobile devices, and the enhancement in technology brings its more uses to make the life of people more productive. The new advancement in tablets, touch phones attracts users in many ways. Companies should also reconstruct the platform to make their campaign successful in marketing mobile services.
4. **Visual Marketing:** The new invention in marketing strategies makes companies more dominant in the field of digitalization. Many new trends are used to involve digitalization in every field as slide share, Instagram, and other websites, and many applications are used for making things valid for customers.
5. **Integrated Marketing:** To make all communication and messaging digitally in integrated form, many companies use 'integrated marketing' to make a hybrid of all activities unified across the world. All the functions related to production, sales, delivery are integrated to make a competent product. Google+ is an invention of Google for catching all patterns and signals in an integrated way.
6. **Content Marketing:** Most of the brands are adopting this marketing strategy because of its importance in 'social media 'and 'multimedia.' This can also be known as a blogging because a blogger is also a content writer. The use of content writing and how 'digital marketing' underlies it somehow not comprehended by many companies. In today's digital field companies need to understand thoroughly the new trends occurred and how it helps in making their brand more powerful (Figure 4.3).

FIGURE 4.3 Types of marketing strategies.

4.6.3 EFFECT ON CONSUMER BEHAVIOR

4.6.3.1 AI PATTERNS USE OF PREDICTIVE ANALYTICS IN CONSUMER BEHAVIOR

To correlating the behavior of customers with their characteristics, predictive models are being used to pump up. These analytics provides recommendations about what type of strategies should follow and how the target audience and their objectives using it got to achieve a particular goal. These analytics helps to find the purchase behavior of customers, what type of trends they follow, or what should be the better strategies to know the changing behavior in purchase patterns. It works through identifying the patterns, extracted the useful data which matches the customer needs than to make validation in models or any modification, 'data mining' techniques are used. Validations of these models are based on those sets of data that choose through identification. The final step is how to apply these models in decisions of business, how applications are used for refining the outcome in a better way.

4.6.3.2 USE OF CHATBOTS

"AI-powered chatbots" are the best example as it provides an instant messaging platform to users for making communication better. 'Zomato' a food application that provides an instant chatting box while food delivery and after if they are not able to satisfy their customer fully. The use of AI chatbots makes companies a unique identity.

4.6.3.3 ADVANTAGES OF AI TO CUSTOMER AND MARKETERS (TABLE 4.1)

TABLE 4.1 Comparison between AI in Customer and Marketer

Customer	Marketer
• Provide 24×7 availability. • The use of AI easily retains the customer's information and reduces the need for interaction again and again. • The time factor becomes reluctant as AI helps to treat many customers simultaneously. • Searching patterns of consumers becomes easy because AI provides many keywords related to customer trends in the suggestion box.	• Promotion campaigns and profits of marketers boost up with the use of AI. • Provide error detection strategies. • 'Medium of choice' is used to provide the right information to the right prospect. • Provide a platform for making the brand successful and prospectively manage consumers. • Making customer relationships for a long time. • Ability to find what consumers want.

4.6.3.4 LIMITATION OF AI

- The available data is limited concerning AI.
- The privacy-related to the customer is not looking attentively.
- AI algorithm is considered time-consuming. It is a difficult task to implement its techniques.
- The maintenance cost of AI adopting systems is high comparing to manual tasks.

4.6.4 HUMAN RESOURCE

Human resource is a department where the use of AI is at the peak. In the new digital revolution where there is a need for resources to get simplified, the use of data analytics, the use of chatbots, 'cloud computing' are in support. In the process of 'recruitment,' to take interviews, to make the list of screening all works related to human resource management are done through AI solutions.

4.6.4.1 ROLE OF AI

1. **Recruitment Process:** The recruitment process, in most companies, is being done by the use of AI. For example, big companies like

'Facebook' 'SAT' mostly use the various techniques of ML to hire a perspective person. With the use of AI, the recruiter quickly performs his/her duty, and the selected candidate easily gets the response of the recruiter. In many exams, the optical mark reader sheets used, which are easily evaluated with the help of AI and other chatbox systems, mostly used to play the best role.

2. **Process Related to Screening/Interview:** To make the process of scheduling the interviews many organization uses the software called 'Amy,' 'Clara.' This software is for meetings related to works or interviews scheduled. Some screening patterns are in the form of speech or verbal reasoning, which are automated with the help of AI solutions.
3. **Manual Works Reduce:** HR is a department that plays in a multitasking form, and with the use of AI techniques, the workload gets reduced and enhances the effectiveness in the HR department by providing a platform of solutions.
4. **Selection:** The selection criteria are to select the right candidate for the right job is easily happened with AI, and the time is going to be very short. AI solutions help to select a suitable candidate for an organization.
5. **Transparency:** The favoritism word is almost reduced in organizations as AI can analyze the curriculum vanity of the particular candidate who is suitable and performs the criteria. It will increase transparency in job centers and organizations.
6. **Efficiency Advancement:** As the use of robots in many organization which performs robotics tasks like filling reports, identify the useful data accounting tasks which increase the efficiency in the work tasks (Table 4.2).

4.6.5 IT SECTOR

The keyword which replaces the traditional way of methods of computing is AI. Mostly in all sectors related to finance, healthcare, industries are transformed in the "blink of an eye." There are positive impacts on AI in information technology sectors as it is already part of this. 'Computer science' born the word 'AI.' All the work related to data transforming, pc related tasks, varieties of software are performed by AI easily. The transformations of old computers into 'intelligent machine models' are a task of AI, which is not possible without the brain of a human being. The 'expert system' is the creation of AI solutions and makes stimulations equal to humans.

TABLE 4.2 Various Benefits and Challenges of the Use of AI in HR

Advantages	Disadvantages
• Employee retention rates are easily predicted. • Maintenance of the workforce and the staff burden gets reduced with AI solutions. • The outcomes are easily obtained with high accuracy. • The biased behavior of humans is minimized, and decision tasks get the support of AI. • It helps to make available the right person for the right job and eliminating the wrong candidates.	• The difficult task for a new learner to study about AI tools. • The right persons for getting useful tools/solutions in the implementation stage get hectic. • The implementation cost is high as difficult to manage for small-scale firms. • Decision-making is a crucial task, as the wrong information can become a hurdle for the human resource.

1. **System Securing:** It builds the securing system as the security of data is a big important task. It stores an ocean of data related to the customer, any private firms, or any type of organization with a high-security level. The use of AI algorithms gratefully performs its tasks as it easily identifies the loopholes or detects errors and data breaches.
2. **Advancing Productivity:** The coding works or aids programmers while overcoming the problem of bugging, the AI models created for advancing efficiency, and it provides suggestions like 'bug-free coding.' It helps to increase the level of productivity.

4.6.6 BANKING SECTOR

Bank Fraud is the means of using illegal activities that give the wrong direction of obtaining money or assets from institutions that provide financial help. There can be fraud like online money transfer fraud, credit fraud, loan fraud, customer on boarding fraud, kiting of checks, some forged malpractices. According to a new "digital tracker," in 2018, \$4.4 trillion losses are due to fraudulent tasks, and there is an increase of 680% in phone application fraud activities from 2015 to 2018. It is observed that one-quarter of malware practices target financial institutions badly.

In the old days of banking, manual procedures, and static rules that are set on specific and static criteria for flagging suspicious acts have been used. Now ML is a technique that simplifies this complex task of preventing frauds and provides better solutions and models that can be used by banking institutions.

4.6.6.1 ROLE OF AI IN THE BANKING SECTOR

The role of AI for mitigating the risk of financial and detecting fraud interpret and improve the goodwill of a particular organization. As the known supervised patterns choose by AI, it also helps to find out any unknown patterns or behavior of transactions and identify the fraudulent tactics of fraudsters. The false-positive situations where all true transactions seem to be a fraud and in false-negative where all true intimidations have been missed are decreased to some extent by AI models. The maximum operations of banks now can be operating by the use of AI analytics [9].

The application of AI detecting fraud helps to review the daily routine activities of clients like surfing or their work history and to find if any money transfer transactions are sync with their original accounts. The AI models prevent and update their data mining techniques and review automatically if any financial fraud is there.

4.6.6.2 SOFTWARE/APPROACHES USED IN BANKING SECTOR

1. **Teradata [10]:** "Think big analytics" a company that provides a platform of intelligence for detecting frauds. They claim that their machine-learning model can enhance efficiency by reducing and detecting fraudulent activities. As there is a bank called "Danske" in Nordics which deals with AI that provide by Teradata. The engine works drastically and identifies the tens of thousands of characteristics regarding online banking transactions providing whether these transactions are fraudulent or fraud-free.

   In today's world, where banks are getting digitalized day per day, fraudulent activities also drastically take place. To resolve these issues, Teradata helps Danske bank to increase its efficiency by reducing fraud activities by 50%. Teradata team analyzes the dataset with the use of ML models that paves the way for better profits widespread in the business. The machine-learning model provides techniques that identify the error in millions of transactions in a single year. There is a layer that works as an interpreter on the upper level of the learning engine. The solutions of Teradata engines reduce false positives by 60% and target to reduce up to 80%.

This is the first bank that implements the anti-fraud program and using AI in a better way. A real-time solution is to be needed for stopping fraud-related activities in online money transfers, credit cards, and other payment tasks.

2 **Feedzai:** It is software used for detecting anomalies in various transactions and providing a solution for money laundering problems and fraud-related acts. As digitization at its peak, fraudsters get maximum opportunities to exploit banking operations, but Feedzai provides end-to-end solutions to tackle this problem.

As in the account opening process in banks, all the KYC norms and verification of clients' details and biometrics and face recognition reduces the friction for the customer and increases the rate of acceptance to some extent. This type of software pays attention and analyzes if any abnormal way of the pattern from normal patterns then may review it or gives an alert to ML tools which are used to determine a fraud if any. It applies behavioral analytics for the customer to keep safe from accounts takeover fraudulent acts. It provides anti-money laundering solutions for automated tools of AI technology [17].

3 **Predictive Analytics:** Predictive solutions are useful to detect sophisticated fraud in various applications of mobile banking, for payment services. It works on the unique method of data analysis in which the pre-trained algorithm measures the risk level of fraud. Prescriptive analytics adopts a way of predictions and provide references for how to tackle fraudulent tasks. From the perspectives of banks, when particular transactions are labeled by banking experts and find some transactions are legitimate and some are fraudulent, then through ML models, the suitable method of recognizing fraud in suspicious transactions can be implemented. There is data visor, which is an AI solution based on predictive analytics:

i. **Data Visor:** The creative writer of data visor [11] is Christopher Watkins. Many companies such as Alibaba, pint rest, and yelp adopted AI Data visor solutions against fraudulent attacks. Data Visor's main aim is to build trust at the online platform. The big team of ML and security experts formed this solution to make available the proven application as it is the unsupervised ML solution though it does not need any trained data for detecting crime reports. It helps to create the cluster of accounts that are linked and take a decision for a time-saving factor through identifying the latest trends of fraud with dashboards.

4.6.6.3 HOW AI HELPS TO TACKLE FRAUD?

According to Adjaoute, 'THE use of AI for fraud prevention is not a new concept, and it's not science fiction.' Many companies like MasterCard and RBS World Pay adopted AI to find any fraudulent patterns and make cards fraud-free to some extent. Banking is a sector in which people's trust builds up, and to promote the area of detecting malpractices, fraud, and illicit banks improve their technology with AI. In today's scenario, the new experiments have been done for managing risk related task that gives way towards the prevention of fraud. According to the "Artificial Innovation playbook," it is reviewed by 63.6% of financial institutions that AI is the best and effective way to curb fraud before it happens. AI helps to find the areas of crime, fraud, cyber fraud, and money laundering effectively [12]; Bank info security [13], security and fraud [14].

4.7 INDUSTRIAL AI

In the industrial sector, where the use of AI was limited earlier, nowadays its applications are being used with the help of creating and deploying new algorithms of ML. Productivity growth has been impacted by its use positively 1. Industrial 4.0 is a change that is an integrated form where the new trends of technology, big data analytics the use of the internet are efficient. The Industrial AI ecosystem is used for a flexible system.

The major elements of industrial AI are 'analytics technology,' 'big data,' 'cybersecurity,' 'domain,' and 'evidence.' The analytics are the core part of technology as they provide a platform of 'value.' If a particular element creates some value the analytics technology helps to bring it. The other element, big data and cyber, provides useful information from the source. It measures the trends of information from various sources and develops the 'industrial AI.' The domain is the element that helps to majorly understand the major problem if there any issue with the system and provides the right solution for this problem. It comprehends the parameters and their uses. The element called evidence which is used in industrial for interpretation of data; the validation of models used and improves the trends to bring more accuracy [15].

4.7.1 *THE ECOSYSTEM OF INDUSTRIAL AI*

The ecosystem in an industrial area where AI performs includes various technologies. These can be 'data,' 'analytic,' 'platform,' 'operational.' These are the major techniques of AI that are used for converting the data, the cybersecurity parts in industries, and the configuration of useful information. Data analytics is for transferring the information from the source to the analysis point. It includes the '3 B' issues, which are background, broken, and bad.

Analytics techniques are used to make the conversion of sensory data to useful data. These are mainly used for predicting the value of the health of the assets. It improvises productivity in a better way.

Platform techniques are used to architect the hardware for analyzing the data and storage capacity formation. The stand-alone and cloud are found to be the main platforms for advancing useful information.

Operational techniques provide a platform for making decisions based on the extracted information. The performance can be optimized through machine-to-machine models. It creates 'self-aware, predictions, configuration, and comparison.'

4.8 CONCLUSION

AI is in a hot spot these days. Countries are taking it seriously and, of course, for the right reasons. AI paves the way for a better and conducive environment in various fields like marketing, accounting, banking, and so many sectors. Development in the field of AI can be a great help to human efforts. Adoption of AI should be embraced, and efforts to increase its potential can do wonders in such a field workload is huge and include an ocean of data. The use of AI in many fields can be more enhanced with new patterns, and there will be a successful attempt to make the use of AI in humans' life.

KEYWORDS

- **algorithm**
- **artificial intelligence**
- **data visor**
- **Feedzai**
- **graphic processing unit**
- **machine learning**

REFERENCES

1. Rupali, M., & Amit, P. (2017). A review paper on general concepts of "artificial intelligence and machine learning." *International Advanced Research Journal in Science, Engineering, and Technology, 4.*
2. Nilsson, N. J., (1998). *Artificial Intelligence: A New Synthesis*. Elsevier. Ibid.
3. Grant, A. W., Seruwagi, L. Z., & Dodd, M. S., (2011). *Artificial Intelligence Through the Eyes of the Public.* https://digitalcommons.wpi.edu/iqp-all/153 (accessed on 1 December 2020).
4. Wierenga, B., (2010). Marketing and artificial intelligence: Great opportunities, reluctant partners. *Marketing Intelligent Systems Using Soft Computing: Managerial and Research Applications, 258*, 1–8.
5. Matsatsinis, N. F., & Siskos, Y., (2002). *Intelligent Support Systems for Marketing Decisions*. Norwell, MA, USA: Kluwer Academic Publishers.
6. Phillips-Wren, G., Jain, L. C., & Ichalkaranje, N., (2008). *Intelligent Decision Making: An AI Approach.* Spring Publishing Company.
7. Crunk, J., & North, M. M., (2007). Decision support system and AI technologies in aid of information-based marketing. *International Management Review, 3*(2), 61–86.
8. Thiraviyam, T., (2018). Artificial Intelligence Marketing. *International Journal of Recent Research Aspects, 19*(4), 449–452.
9. Net Guru, (2019). In: Dyzama, M., (ed.), *Fraud Detection with Machine Learning: How Banks and Financial Institutions Leverage AI.* Retrieved from: https://www.netguru.com/blog/fraud-detection-with-machine-learning-how-banks-and-financial-institutions-leverage-ai (accessed on 1 December 2020).
10. Teradata, (2017). *Danske Bank and Teradata Implement AI Engine that Monitors Fraud in Real Time.* Retrieved from: https://www.teradata.com/Press-Releases/2017/Danske-Bank-and-Teradata-Implement-AI.
11. Data Visor, (2019). DataVisor in 2019: The Year in Review. Retrieved from: https://medium.com/datavisor/datavisor-in-2019-the-year-in-reviewb7c9b810b9c8 (accessed on 1 December 2020).

12. Harvard Business Review, (2018). *The Risks and Benefits of using AI to Detect Crime.* Retrieved from: hbr.org/2018/08/the-risks-and-benefits-of-using-ai-to-detect-crime (accessed on 1 December 2020).
13. Bank info Security, (2014). *How Artificial Intelligence Prevent Fraud.* Retrieved from: https://www.bankinfosecurity.com/akli-adjaoute-a-6547 (accessed on 1 December 2020).
14. Security and Fraud, (2019). *AI Helps Businesses Get Smarter about Fraud.* Retrieved from: https://www.pymnts.com/news/security-and-risk/2019/fraud-prevention-artificial-intelligence-innovation/ (accessed on 1 December 2020).
15. Lee, J., Davari, H., Singh, J., & Pandhare, V., (2018). Industrial artificial intelligence for industry 4.0-based manufacturing systems. *Manufacturing Letters, 18*, 20–23.
16. Ace Cloud Hosting, (2018). *How Artificial Intelligence Impacts on Accounting Industry.* Retrieved from: https://www.acecloudhosting.com/blog/artificial-intelligence-impact-accounting/ (accessed on 1 December 2020).
17. Feedzai, (n.d). *Feedzai for Retail Ranking.* Retrieved from: https://feedzai.com/solutions/retail-banks/ (accessed on 1 December 2020).
18. Forbes, (2018). *Here Are Three Factors That Accelerate the Rise of Artificial Intelligence.* Retrieved from: https://www.forbes.com/sites/janakirammsv/2018/05/27/here-are-three-factors-that-accelerate-the-rise-of-artificial-intelligence/#3f405f06add9 (accessed on 1 December 2020).
19. Random Forest, (2019). *Random Forest.* Retrieved from: en.wikipedia.org/wiki/Random-forest (accessed on 1 December 2020).
20. Simple AI, (2017). *Logistic Regression-Intro to Machine Learning.* Retrieved from: https://medium.com/simple-ai/logistic-regression-intro-to-machine-learning-7-ba18ab305b24 (accessed on 1 December 2020).
21. Watkins, C. (2019). Retrieved from: https://www.datavisor.com/blog/best-ai-based-solution-for-fraud-management-2019-ai-breakthrough-awards/ (accessed on 1 December 2020).

CHAPTER 5

REVOLUTION THROUGH ARTIFICIAL INTELLIGENCE IN CONTEXT TO FINANCIAL SERVICES

SRISHTY MEHRA[1] and GURMEET SINGH[2]

[1]*University School of Financial Studies, Guru Nanak Dev University, Punjab, India, E-mail: srishty012@gmail.com*

[2]*Head of Graduate School of Business, University of South Pacific Statham Campus, Fiji*

ABSTRACT

The human aspect of learning new things and the problem-solving in arising situations can now be done by the new age computers with machine intelligence known as "smart computers," while, the machine intelligence also known as Artificial intelligence covers diversified fields, including medical, engineering, finance, etc. the potential of AI is still unknown as it is not completely explored, but there are expectations that it will change the way of machine operations and mark the beginning of the new modern era of humanity. While in the current age of technology, all the banks and non-banking financial companies rely heavily upon the internet and computers for providing financial services to their clients hence the introduction of AI in the financial service sector will be a game changer by facilitating innovation of new financial service products for the people. In recent times, PayPal, a company dealing with online funds transfer around the world, has used AI system for strengthening the cybersecurity for online transactions resulting more reliability of customers as well as adding more customers on the payment network [4]. Although AI offers some promising outcomes, but some key challenges, includes the need for high-end technology for the cost-effective integration of AI with financial services and while as lack of AI development at commercial scale and limited integration of AI in educational courses resulting a handful of people well qualified and experienced in the field of AI [5]. Also, lack of understanding about AI in the public domain will cause suspicions in the minds of governments posing risks of unhealthy government regulations in the future. Another problem relates to independent working processes of AI in the future leading to the complete loss of human control over AI and may even establish one AI bot to other AI Bot communication as a similar incident was observed in case of Facebook where two AI Bots start communicating in a language not understood by humans.

5.1 INTRODUCTION

Today artificial intelligence (AI) is considered a hot topic which can be seen on any present issues paper, magazine or diary, and the internet is flooded with videos, blogs, and articles on various issues regarding the manner the 'robots are dominating' and how this baffling innovation is the greatest hazard to humankind since the atomic bomb was developed. In the meantime, the organizations making AI applications synthesizing fabulous cases of innovation, clarifying how it will articulate revolution in the livings of people while mudding any genuine incentive in a fog of showcasing exaggeration. In addition, afterward, there is simply the real innovation—a delusion of arithmetic, information, and PCs—that gives off an impression of being a miracle technology to anybody outside of the designer world. Hence it wouldn't be surprised when we talk about perplexity in the minds of business houses about the fascinating accomplishments of AI. So what precisely is meant by AI? What miracles can it perform?

In what capacity can it advantage one's business? What should be the inception? The substantial inquiries yet unanswered needs to be explored in the sea depth of AI applications [6]. Man-made consciousness, in its broadest sense, will fundamentally affect how we work together. Of that, there is no uncertainty. It will change how we decide, and it will empower new plans of action to be made, and it will enable us to do things that we at no other time thought conceivable. Be that as it may, it will likewise displace the work as of now being finished numerous information peasants and will unequally compensate the individuals embracing AI adequately as early adopters.

We can see the hints of the start of current machine learning (ML) as an endeavor to characterize the traditional thinkers' arrangement of the human idea as a symbolic framework. Be that as it may, the field of ML was not officially settled until 1956. In 1956, a gathering "ML" was held without precedent for Hanover, New Hampshire, at Dartmouth College. Intellectual researcher Marvin Minsky at MIT and different researchers taking an interest in the gathering was very hopeful about the eventual fate of ML. As Minsky expressed in his book "simulated intelligence: The Wild Search for AI": "In age, the issue of man-made consciousness creation will be comprehended at a critical level." One of the most significant visionaries and theoreticians was Alan Turing (1912–1954): in 1936, the British mathematician demonstrated that an all-inclusive mini-computer-presently is known as the Turing machine-is conceivable. Turing's focal understanding is that such a machine is equipped for taking care of any issue as long as it tends to be spoken to and understood by a calculation. Moved to human knowledge, this implies if intellectual procedures can be calculated, then it can be separated into limited well-characterized individual advances they can be executed on one machine is both an enormous chance and an inauspicious danger enveloped with a puzzling heap of calculations and language. Yet, this innovative unrest isn't something that will occur later on; this isn't some hypothetical exercise that will concern a couple of organizations [7]. Computerized resourcing is being utilized today in organizations to expand, enhance, and alter the manner of their endeavors. Open-minded administrators are now figuring out the approaches by which AI can augment the worth of their organizations, looking to see all the various kinds of AI, and also figuring out the measures to moderate the dangers that would be brought by it. Those endeavors are covered up or stayed quiet by their agitators in a significant number, in light of the fact they didn't require the utilization of AI in their commodities or administrations to be known generally or because they would prefer not to bestow the upper hand given by it. Administrators have been persistently confronted with diligent tests that need to get holdings with AI in discovering all requisite data without depending upon the whimsical editorials, tuning in to seller overstatement, or attempting to get calculations.

Man-made intelligence is solidly in the field of 'cognizant questions'—we realize that we lack requisite information. Individuals, for the most part, experience AI-first as shoppers. All our cell phones approach complex AI, irrespective of the fact is it Cortana, Siri, or Google Assistant (regardless of whether that is Siri, Cortana, or Google Assistant). Remarkable homes are presently AI-empowered via Google Home and Amazon's Alexa. They, as far as anyone knows, make our lives simpler to sort out, and for the most part, they do a quite great job of it. In any case, their utilization of AI is, in reality, entirely constrained. A large portion of them depends on their capacity to transform their discourses into alphabets, and afterward, those alphabets into significance. *After marking the milestone*, the remainder of that errand is truly accepted mechanization; discover the climate estimate, get train schedules, playing a melody. In addition, despite the fact that the discourse acknowledgment and common language understanding abilities are extremely shrewd in what they accomplish, AI is far beyond that, particularly in the realm of business.

The expression "robo-guide" was inconceivable only five years prior, yet it is presently ordinary in the monetary scene. The term is deluding and doesn't include robots by any means. Alternatively, maybe, robo-counsels (organizations, for example, Betterment, The researcher front, and others) are calculations worked to align a monetary portfolio to the objectives and hazard resilience of the user [8]. The framework at that point adjusts to changes in the client's objectives and continuous changes in the market, pointing consistently to locate the best fit for the client's unique objectives. Robo-counsels have increased critical footing with millennial purchasers who needn't bother with a physical consultant to feel good contributing, and who are less ready to approve the expenses paid to human counselors.

Man-made brainpower can peruse a great many lawful arrangements in minutes while concentrating all relevant, helpful data appertaining to them; it has the capability of distinguishing carcinogenic tumors with more prominent exactness as compared to the human radiologists; it also has the ability to recognize fake charge card conduct prior it occurs; it also can operate vehicles without chauffeurs; it also can manage server farms more effectively as compared to people; it also can anticipate when clients (including representatives) are traveling to forsake one and, in particular, it also can learn by itself and develop dependent upon their encounters. Be that as it may, until business administrators comprehend what AI comprehends or appreciate, in the straightforward abundant nitty-gritty, and how can it support one's business, they will never arrive at their maximum capacities. Those having the foreknowledge of its utilization and reprimanding AI advances are the individuals that should be called for recognition about what miracles it can perform and comprehend what actions they have to perform for making things move in the requisite direction [9].

5.2 NEED AND IMPORTANCE

We need AI because the work that we have to do is expanding every day. Therefore, it's a smart thought to robotize the normal work. This spares the labor of the association and expands the efficiency. Moreover, through this AI, the organization can likewise get the gifted people for the improvement of the organization. In addition, the organizations today believe that they need to motorize all the standard and routine work. What's more, they want to computerize those customary works through the basic program because, with the advancement of information science, robotization turns out to be increasingly normal. The use of this AI is significantly observed at the site visit gateway.

5.2.1 RESPONSIVE MACHINES

A genuine case of these Reactive machines is deep blue. What's more, this is typically applied to the Chess Board. This can distinguish the bits of the chessboard and make forecasts [10]. In any case, the downside is that it doesn't have memory. That is to say that it can't utilize the past encounters to foresee what's to come. Therefore, it just utilizes the present circumstance and just moves the pawn. It is planned for the use of a little circumstance. The circumstance could be taken care of inside a minute itself.

5.2.2 CONSTRAINED MEMORY

This kind of AI is applied where there is a requirement for memory. A few machines chip away at the premise of the experience. It implies they the data of something very similar that occurred previously and does the present work likewise [11]. This kind of the AI lion's share is utilized in the territories of self-driving vehicles. It utilizes its memory to act in the circumstances like car accidents.

5.2.3 HYPOTHESIS OF BRAIN

This sort of AI alludes to the comprehension of different practices and take every necessary step as per It incorporates the inclination. The aim moves of the other individual [12]. All in all, this sort of AI still doesn't exist. Also, today the vast majority of the researcher's product endeavoring to get it. Like the above kind, this sort of AI framework needs to bring for all intents and purposes. They believe that these frameworks ought to have mindfulness and must have the ability to comprehend the sentiment of others.

5.2.4 AUTOMATION OF MANUAL WORK

AI robotizes monotonous research and revelation by dint of information. In all cases, AI is unique concerning equipment guided, automated mechanization [13]. Instead of mechanizing the manual undertakings, Also AI performs visits, tremendous-volume, automated errands conscientiously without any weariness [14]. For this sort of computerization, a human appeal is as yet fundamental to raise the groundwork and pose correct inquiries.

5.2.5 KNOWLEDGE ADDITION

Man-made intelligence adds knowledge to existing items. As a rule, AI can't be made to be available in the market as an individual application. Alternatively, maybe, technologies you being now used shall be improved with AI capabilities, considerably like Siri, which was supplemented being a component of Apple techs in other ages of this era.

Automation, conversational platforms, savvy machines, and bots have been incorporated with abundant information for improvement resulting in plentiful upgrades in the work settings and at home, from the knowledge of security to speculation investigation.

5.2.6 EXPERTISE

Man-made intelligence adjusts through dynamic learning calculations for giving a chance to the information to execute programming. Computer-based intelligence unearths frameworks and consistencies among knowledge so that most of the calculations get the requisite expertise [16]. Then the calculations are flipped into the classifiers or gauges. Along these lines, similar is the situation when calculations can show the tactics and moves of playing chess, and it can also showcase by itself that what items to prescribe next on the internet. What's further, the models have self-learning systems that adjust it when fed with innovative information [17]. Back proliferation is a kind of AI system which sets up the model to modify it by preparations with included information when the principal answer isn't exactly right.

5.2.7 ERROR AND FRAUD DETECTION

Join progressively available registering scientists, web winding up more regularly utilized, and a broad gauge of important organization's data being locked up on the internet, and you have an "impeccable tempest" for information security hazard. During past monetary extortion, location groundwork's depended eagerly upon intricate along with powerful compositions of fundamentals, and present-day misrepresentation identification goes past afterward the scheduling of hazardous elements—it energetically grasps and regulates to new genuine (or potential) surveillance dangers [18]. This is the spot of AI in money for extortion—however, akin principles stay steady for more information surveillance affairs. Utilizing AI, groundwork's can recognize extraordinary practices ("oddities") or exercises and banner them for surveillance associations. An analysis for those groundworks is to avert from bogus-positives-circumstances whereby "dangers" are addressed in order so that the specialists never peril in either circumstance [19]. Here at Tech Emergence, the researcher interview the analyst about six misrepresentations and security AI administrators, every one of whom appears to be persuaded that given the inestimably high number of ways that security can be broken, truly "learning" frameworks will be a need in the five to ten years ahead.

5.3 VALUE ADDITIONS TO FINTECH CORPORATES

Financial organizations occurred to be the prompt adopters of the centralized server PC, social databases, and have energetically anticipated the following degree of computational capacity. Inorganic brilliance aids Fintech organizations in the tackling of individual's issues through expanding effectiveness. Computerized reasoning (AI) enhances outcomes through the application of techniques derived from parts of Human Brilliance appearing in past individual's scope. The most recent computational weapons contest among two decades has altered the FinTech organizations. Advances in the form of Machine Self Enlightenment, AI, Neurological Networks, Massive Information Analytics, transformative calculations, along with more significant empowering PCs for crunching gigantic fluctuated, various profound data sets in comparison to any time in recent memory.

5.3.1 PRECISE DECISION-MAKING

Information driven administration choices at dropped price drove to an alternate expression of the board, whereby protection pioneers along with future financial specialists would pose the correct inquiries to machines instead of to human specialists. Machines will, at that point, investigate the information and will concoct the suggested outcomes [20], which can support pioneers and their subordinates to take a better choice.

5.3.2 QUANTITATIVE AND FINANCIAL ANALYSIS

With enormous volumes of quantitative and precise verifiable information, ML would now be able to be utilized in money related examination. ML is now being utilized in the account for a portfolio the board, algorithmic exchanging, advance guaranteeing, and misrepresentation identification [21]. However, future utilization of ML in the account will incorporate Chatbots and other conversational interfaces for security, client care, and notion examination.

5.3.3 COMPUTERIZED CUSTOMER SUPPORT

Clients, confronting frameworks, for example, content visits, voice frameworks, or chatbots, can convey human-like client assistance or master counsel involvement with minimal effort. How Mechanization or Chatbots can benefit Fintech organizations set aside their cash and time are talked about as follows: Chatbots as your Finance Assistant.

5.3.4 REMOVAL MANUAL DATA ENTRY

Copy and off-base information are the absolute most concerning issues looked at by the organizations today. Prescient displaying calculations and ML can fundamentally keep away from any blunders brought about by manual information passage [23]. ML projects improve these procedures by utilizing the found information. In this manner, the workers can use a similar time for doing errands that increase the value of the business.

5.3.5 INSURANCE UNDERWRITING

Guaranteeing can be characterized as an optimal activity for AI in the account, but in reality, plenty of stress lies there in the businesses that a huge swath of the endorsing places would be supplanted by machines that prevail nowadays. Notably everywhere institutions (huge traded and banked upon an open market conservated enterprises), AI computations can be worked upon a great many instances of purchaser data (like conjugal status, work, age, and so on…) and financial loaning or conservative outcomes relating to person's fault, the advance reimburse schedule, get in a fender bender, and so on [24]. The fundamental arrangements can be figured out accompanying the computations and constantly streamlined for recognizing sequences that may influence safeguarding and loaning into what would arrive (are consistently rising figures of youngsters in a clear-cut case showing up in fender benders? Were there aggravating paces of default among an appropriate statistic populace in the policies since preceding 15 years?) These results have a colossal unmistakable yield for organizations-yet at present are held for huge institutions having assets for procuring information analysts and the enormous quantities of more significant time interval information for preparing their algorithms [26]. The specialist thought about the AI speculations of protection mammoths like State Farm, Liberty Mutual, and others-in our total article on AI protection applications [26].

5.3.6 EXTORTION DETECTIONS AND CLAIMS MANAGEMENT

Investigation apparatuses gather proof and break down information vital for conviction. Man-made consciousness apparatuses then learn and screen the client's conduct standards to recognize irregularity and cautioning indications of extortion endeavors and rates [27]. Cases the board can be developed utilizing ML procedures in various phases of the case dealing with the system. By utilizing AI and taking care of an immense measure of information in a little timeframe, back up plans can computerize taking care of instrument [28]. It can even affix certain cases; diminish the general preparing time and the taking care of expenses while upgrading client experience. These calculations distinguish designs in the information to help perceive fake cases simultaneously. With their self-learning capacities, AI frameworks would then be able to adjust to new, unfamiliar cases and further upgrade the discovery after some time.

5.3.7 AS WEAPON IN BAD TIMES

Since the worldwide monetary emergency in 2008 and guidelines from Basel to Sarbanes-Oxley, cost weights have been collecting on banks. It has additionally made tremendous potential for disturbance. Presently the inquiry is whether the worm will turn or if fin techs will progress toward becoming business sector pioneers. Today, fintech and banks are dominatingly correlative. It stays to be perceived how rapidly the enormous players will understand the chances of digitalization—that is, not just supplanting simple procedures with computerized, yet finding new potential in datasets and AI.

The greatest test is likely social. Computer-based intelligence needs a methodology of 'flop quick, bomb brisk,' however, banks still think that it's difficult to acknowledge disappointment [39]. With RPA and AI, the money related world presently has an approach to allow workers to begin this social change. Go on, set out to venture out.

5.4 CASE STUDIES

5.4.1 QUILL

Quill is a computerization stage created by a Chicago-based organization of Narrative Science. Quill helps in money related detailing, an AI innovation that changes over your information into plain English to cause you to comprehend the bits of knowledge in basic words, much the same as a human information examiner would do in your organization [41]. Quill additionally comprehends the most pertinent piece of a report from records and attempts to give the reports in comparative words, sentence length, and structure to make it clear for non-specialized individuals. It spares huge time for the undertakings examining information and making manually written outlines and reports. Account Science has additionally helped to keep money with the age of programmed suspicious movement reports. In this way, Quill is valuable for money related foundations in diminishing dangers and looking after consistence.

5.4.2 ROYAL BANK OF SCOTLAND

In 2017, Royal Bank of Scotland acquainted a chatbot with assistance to answer a predetermined number of inquiries from clients. The chatbot, called 'Luvo,' employments IBM Watson's Conversation capacity to interface with clients who are utilizing the bank's site or application. It was trialed for almost a year with inward staff who oversees associations with SMEs before being discharged to a small number of outer clients [40]. At the hour of composing, it can answer just ten characterized questions, for example, 'I've lost my bank card,' 'I've bolted my PIN' furthermore, 'I'd like to arrange a card-per user.' Luvo exhibits the mindful methodology that endeavors are taking to executing chatbots. After some time, however, as the framework gains from each cooperation it has, the bank will enable it to deal with increasingly complex inquiries, work in more noteworthy personalization and utilize prescient investigation to distinguish issues for clients and afterward to prescribe the most suitable activity. The key goal is to save more opportunity for the human client support specialists to handle the trickier inquiries that clients may have.

5.4.3 SEB: THE LEADING NORDIC CORPORATE BANK

In a comparative two-advance usage, SEB, probably the biggest bank, conveyed IPsoft's Amelia programming first in their inner IT Service Desk, and at that point, to their 1 million clients. During the initial three weeks of the preliminary, the chatbot-cum-symbol addressed the staff's IT issues, settling around half of the 4000 questions without human cooperation. During 2017, it was taken off to help oversee cooperations with the bank's clients [42] (and named 'Aida'). Three procedures were at first picked as competitors: giving data on the most proficient method to turn into a client, requesting an electronic ID, and clarifying step-by-step instructions to do cross-fringe installments. Even though Amelia gains from authentic collaborations and utilizations slant examination, it is likewise ready to pursue characterized work process ways to guarantee consistency with banking guidelines. While most chatbots can be portrayed as probabilistic, giving likely outcomes, Amelia can be viewed as a deterministic framework because, when it has worked out the expectation, it can, where conceivable, do the required activities on the venture frameworks (along these lines that RPA devices would). IPsoft is a US-based organization, and this was the first non-English arrangement of their arrangement. From a framework perspective, SEB chose to introduce the Amelia innovation all alone servers as opposed to as a cloud arrangement. This was expected to worry over the consistency and lawful worries that a cloud organization raises. For IPsoft's money related to administration clients, this is the most widely recognized methodology. The sending of Amelia is specially lined up with the bank's by and large procedure, which incorporates the line "SEB will concentrate on giving the main client experience, putting resources into advanced interfaces and robotized forms" [43].

5.4.4 1-800-FLOWERS COMPANY

The US bloom conveyance administration 1-800-Flowers sent a straight-forward chatbot that enabled clients to place arranges through Facebook Messenger. It is a straight, choice tree framework with an NLU visit interface that took around three months to create and test. This makes it somewhat obliged in what it can do, yet extra usefulness and multifaceted nature are being taken off. Following the underlying two months of activity, 70% of the requests being put through the chatbot were from new clients and were ruled by 'recent college grads' who will, in general as of now, be overwhelming clients of Facebook Messenger [44]. Just as taking requests, the chatbot can guide a client to a human specialist, of which there might be up to 3,500 at any one time. They have moreover executed a joining with Amazon Alexa, and an attendant service that is controlled by IBM Watson. Together, these computerized activities have pulled in a huge number of new clients to the brand. They likewise give up-to-the-minute conduct information for the organization, which would then be able to impact continuous showcasing action, for example, advancements.

5.4.5 STAPLES COMPANY

Another organization that has utilized IBM Watson to upgrade client commitment is Staples, the stationery retailer. The company has executed a scope for various approaches for individuals to purchase the required items as effectively as could be expected under the circumstances, including Slack, e-mail the versatile application along with (actually) a major red catch. This catch is comparable to Amazon's Alexa in the way that it can comprehend speech directions (even though it needs a physical press for actuating and deactivating it). This portable application can recognize voice directions just as having the option to distinguish items from photos taken. Every one of those channels performs the purchasing procedure as continuous as workable for the client and accordingly has an immediate and constructive sway upon income for the shopkeeper. Just in the case of chatbots, proposal motors also perform another basic AI innovation, which is utilized for upgrading client assistance (and drive income). Amazon and Netflix have the most outstanding proposal motors, and these are profoundly inserted in the typical work process of how the client locks in with the organizations. All the required information—the individual client's purchasing and perusing practices and the recorded information of the various clients' practices—are accessible through those typical communications; that is, there is no extra work that the client needs to do to improve the informational index. Now and again, the suggestion motors will require extra data from the client as well as the business to work adequately. North Face, an attire retailer, has actualized a suggestion motor for their clients who need to purchase coats. In light of IBM Watson, the arrangement, known as XPS, utilizes a chatbot interaction to pose a progression of refining inquiries with the goal that it can coordinate the clients' necessities with the product offering. As per North Face, 60% of the clients navigated to the suggested item.

5.4.6 CLYDESDALE AND YORKSHIRE BANKING GATHERING (CYBG)

Clydesdale and Yorkshire Banking Gathering (CYBG) is a medium-stature bank in the UK, being to contend with RBS, Lloyds, HSBC, and 'Huge Four' of Barclays. The computerized methodology incorporates another present record, investment account, and application bundle, known 'B.' This utilizes AI for helping deal with the client's cash: this enables to you accessible record, and, when opened, would develop into familiar with the examples of utilization so that it can foresee on the off chance that you may come up short on assets in your records, and propose approaches to maintain a strategic distance from superfluous charges of the bank. Now a day's bank guarantees that a record can be opened in 11 minutes. Clients staying away from charges of the bank would result in lowering income; however, this expectation draws in ample new clients that this income misfortune would effectively exceed through advantages of possessing the extra reserves. Virgin Trains 'reimburse/delay' operations have been robotized by the application of AI. After actualizing Celaton's in STREAM AI programming to order inbound messages (see coming areas), the train working organization employed the same programming to provide a sans individual's interaction to clients to consequents guarantee discounts for deferred trains. The famous Asian life insurance provider AIA has executed a wider scope of AI activities that the cumulative effect how would they draw in with the clients, including knowledge age from forthcoming clients, improved evaluation of client requirements, 24×7 hours chatbots for entertaining online inquiries, inbound call taking care of by NLU-established frameworks, upgraded consistency of offers, customized estimating, aggressive guaranteeing along with an increased exhortation and suggestion motor.

A few organizations are incorporating AI with the center of their client applications. Under Armor, a games attire organization that has an arrangement of wellness applications, utilizes AI among those applications to give preparing programs and proposals to the clients. The AI grasps information from an assortment of origins, counting the client's different applications, dietary databases, physiological information, social information, and results from different clients with comparative profiles and targets. At that point, it gives customized nourishment and preparing exhortation that moreover considers the hour of the day and the climate.

5.4.7 PayPal HOLDINGS

PayPal Holding Inc., a renowned US company, performs its business activities around the globe for online payment systems supporting electronic money transfers by overruling the traditional paper methods, including check and money orders. This company is providing 360° payment processing facilities to auction sites, online vendors, and many more commercial benefits, including one-click transactions and password store facilities.

If we consider statistics of the year 2019, PayPal administers in about 202 markets, having about 286 million registered active accounts. PayPal helps its clients to receive, send, and detain funds in about 25 different world currencies.

PayPal has facilitated the people for making electronic financial transactions by allowing them to move funds online between business houses and individuals. By the grace of PayPal services, users are now able to receive and send payments on online websites, including E-Bay, can procure or auctions goods and services, and also enables them to receive or donate donations electronically. It also provides a marvelous facility that PayPal doesn't require accounts as users are free to set currency conversion settings under account settings. From the year leading from 2009 to 2016, PayPal administered accounts for students; authorizing parents for having students account for their champs, and easily transmit money and also provide plastic cards for usage to students. This program also enlightened the path of improvising the spending habits and also became responsible for its activities. However, unfortunately, the facility of the student's account was discontinued effect from August 2016.

5.4.8 CONTEXTUAL ANALYSIS OF BARCLAYS BANK

As of late, Barclays has looked to build up itself among the Five Big banking institutions in the UK among the main speculators in AI. Like most of the banking institutions, Barclays also intend to locate the remarkable financially savvy methods for improvising its administrations via adjusting them for making it simpler for their clients to get through, explore, and Barclays is not distant from everyone else in perceiving the potential AI offers for conveying upon the point, especially recently when the psychological capacities in robots are developing. Accompanying the enormous measure of information which banking institutions, for example, Barclays has, AI that would empower money related establishments to develop such information at record limits and speeds. Moreover, improvement in intellectual abilities concerning AI, the arrangements it would have the option to propose would extend along with the furnishing clients with significantly extra vivid what's more, customized client experience. One of Barclays' first significant strides within the universe of AI returns in the year 2012, dispatch of their web-based banking application 'Pingit.' That application empowered clients of the banking institutions to transmit cash from one person to another through their cell phones and via web-based networking media stages, for example, Twitter. The application was along these lines created to empower clients to make in-shop buys utilizing the app. Increasing interest in AI has all the earmarks of being the general pattern at the Barclays. It has guaranteed that the goals following the venture are to give progressively customized administration for their clients. Just like its internet banking applications, it has utilized AI for creating along with improvement of its capacities among different zones. One zone in which it is looking to utilize AI is in the administration of online client inquiries, utilizing chatbots to address client questions in a practically prompt manner.

Barclays UK is not utilizing this specific kind of innovation by and by yet. Be that as it may, they are running various preliminaries in their South African backup 'Absa.' Everyone at Barclays is unmistakably on the path of empowering more noteworthy utilization of Computer-based intelligence. 'Bird Labs' have been set up by Barclays in the branches where the administration was streamlined. Sessions are held for the clients in these labs to evaluate the utilization of new advanced administrations, for example, online banking. Barclays is continuously working towards introducing robots to supplant backend office errands to achieve effectiveness.

Another bit of innovation that Barclays have put resources into its voice acknowledgment, using which within the arrangement for phone-centered client support and provide better client data safety. Finally, Barclays are putting resources into the utilization of AI In checking client exchanges and misrepresentation prevention. By distinguishing designs in the authentic information the bank holds, AI can distinguish strange practices, for example, tax evasion, illegal exchanges, and security dangers. The primary motivation for Barclays' efforts in the automation field seems, by all accounts, to be two-crease: Right off the bat, Barclays perceive the money-saving advantages that technology will provide arrangements in budgetary administrations.

Considering it straightforwardly, while AI takes more time to grasp than people, AI can achieve a few undertakings more adequately and productively as compared to people falling in the long haul. Barclays has a tremendous measure of information on the company's client's money related to past events and activities. Taking regard of specific undertakings, for example, misrepresentation discovery and avoidance, for instance, information that would consume individual's timings, but the same is not happening nowadays to distinguish and break down, though a robot can perform the same activity or function within few seconds. Besides, the most evident favorable position of utilizing a robot to perform such undertakings is that it shouldn't be compensated. Subsequently, clients are accepting quicker, less expensive, and better benefits at Barclays. Another impetus driving Barclays' interest in AI gives off an impression of being a reaction to new rivalry available. Challenger banks, for example, Fidor, Monzo, and Atom, that are simply innovation-driven, are drawing potential and present clients from Barclays also, the other 'Huge Five' banking institutions. Numerous recent college grads and persons under 40 years of age transact the majority of their banking online. It is viewed as that just on uncommon events, and these aged gatherings approach the branch to the banking institution. As of late, challenger-banking institutions have exploited this enabling them to rule the specialty markets. Monzo's exclusive item is that they are paid ahead of time by MasterCard, so for what reason are there effectively 400,000 clients at Monzo, with a holding up rundown of 25,000? Its innovative marketing is the cautions on cash the board. Focusing upon versatile applications of it, Monzo alarm their clients on their expenditure and use programming to plan spending plans for clients naturally. Even though Monzo is not offering home loans, ISAs, or investment accounts, they can exhort clients' purchase points of the items and naturally convert their cash into those physical commodities. The comfort and customized banking background have prevailed upon numerous clients. The long haul eventual fate of challenger banks, for example, Monzo will bear some significance with the money related segment. In July 2017, Monzo revealed misfortunes of £7 million, and 65 startup organizations regularly need money related help from financial speculators to get by in the early years. Something else, five UK banks are planning to acquire Monzo. Meanwhile, banking institutions, for example, Barclays, are contending with challenger banking institutions via putting resources into the tech-sphere.

5.5 CHALLENGES

Difficulties of AI in the monetary business, surviving, writing, and the above contextual investigation featured various lawful and moral difficulties presented by AI. Researchers have understood that AI projects can separate on the secured attributes of race and sexual orientation. Men get more high paying Google search work ads than ladies a wonder regularly famous do with the term 'white person' disorder in algorithms. Microsoft's chatbot Tayai chirped supremacist and misogynist remarks when an attempt to emulate the dialect examples of a 19-year old American girl. This segment centers on the affairs of inclination and calculations, as it represents the obstacles that may be brought through Big Data, calculations, and AI. Perpetual threats of straightforwardness, as well as protection, will likewise be examined. The inclination of calculations can happen in three phases: info, preparing, and programming. The information can be conceivably one-sided, or the data is lacking for the calculations to settle on reasonable choices. ML envelops the preparation and programming stages. Managing input, banks never again depend on conventional information, for example, the probability of reimbursement, the obligation to income, what's more, advances to-esteem proportions, financial record, and scores of eligible borrowers to survey them. Rather, web look, shopping examples, and internet-based life exercises give banks new information. They incorporate a potential borrower's inclinations on vacations, leisure activities, interests; work-related pursuits; associations; social exercises, and so on. Utilizing non-customary information can open money related approaches to borrowers having inadequate money related records or financial records. Taking, for instance, the 'Tala' application utilizes versatile information to make a money related personality to borrowers in rising nations, e.g., Asia and Africa. Money related character is built up through information, for example, a borrower's security in key networks; associations; arrange decent variety; area consistency; money related exchanges. As indicated by the authors Shivani Siroya, a borrower within an event, 58 contacts is bound to reimburse his/her loan. Tala's reimbursement pace of 90% is impressive. These are, on the whole, reassuring news; however, non-customary information pulls in the issues of information unwavering quality what's more, protection. On the off chance that AI calculations customized that the normal FICO rating of a 'companion' via web-based networking media stages is utilized as an indicator of financial soundness, in what manner can they tell whether the fellowship is authentic or not? Consider the possibility that the 'companions' are the guardians of a borrower. Who has gone about as his/her underwriters in the past as a coincidental occurrence? Foundational inclination might emerge if the AI calculations are modified such that monetarily capable customers are probably going to associate with monetarily capable individuals. Besides, non-conventional information is gathered through plenty of techniques. Borrowers may not know that information gathered on state web searches or buys will be utilized to evaluate their financial soundness. On the off chance that their advance applications are rejected, at that point, the European General Data Protection Regulations 2016/679 (GDPR) may aid hypothesis. The GDPR gained its power in May 2018 regardless of Brexit. The United Kingdom govt. expressed that they would execute an equal system. The GDPR provides buyers the privilege to learn when organizations are settling on robotized choices of any significance about them. The GDPR additionally gives the privilege to challenge computerized choices. In any case, once there any type of human mediation lies in the basic leadership process, the privilege to challenge won't make a difference. It just applies to choices that are robotized completely or halfway through the whole procedure (Articles 2 and 22). As observed before, chatbots and robo-counselors would require human counselors to help with progressively intricate, touchy issues. It is hence impossible that robo-exhortation in monetary administrations will be computerized, surely in the near future. In this manner, as a general rule, the privilege to challenge options underlying the GDPR is improbable to give more assistance to purchasers. The second and third phases, where inclination can happen, appear in preparing and studying. Preparing can appear as regulated learning, unaided learning, or fortification learning. Supervised learning happens when PCs are nourished with the two data sources furthermore, yield. For instance, PCs can recognize credit extortion by gaining from misrepresentation scores produced a rundown of effective exchanges and unlawful exchanges. PCs in regulated learning will be given the two sources of info and choices. In unaided learning, be that as it may, PCs need to settle on their own choices about credit extortion. This is the place possibly, and PCs can settle on one-sided choices relying upon which information the engineers believe are significant. Predisposition can likewise occur in fortification learning. Here, PCs consolidate a few choices to derive the most out of it. The Deep Mind calculation utilizes a variation of the Q-learning calculation to defeat human adversaries in three arising out of seven rounds of Atari. In March 2016, the Alpha Go program of Deep Mind defeated the human boss in Go by four games to one. The program had the option to study solo from its very own mix-ups and self-play. AI is modern to the point that even designers who have made the calculations may not completely grasp how they have advanced. The risk of this is that the calculations may accidentally make a predisposition even where there is no expectation to do as such. Give us a chance to consider two theoretical situations where calculations are utilized to anticipate purchasers' reliability. Calculations that utilize instructive degrees of customers as a model for financial soundness screen their spelling botches on a web look. If instructive levels lie at lower levels in ladies in a specific race, at that point, this can prompt aberrant sex discrimination. Algorithms that screen purchasers' internet spending at rebate stores anticipate that there is a bigger danger of default of advances among many customers. If these markdown stores are excessively situated in ethnic minority networks, once more, this can prompt roundabout discrimination. Supreme Court in the year 2015 the United States held instance of Texas Department of Housing and Community Affairs v. Comprehensive Networks Plan Inc. that lodging and loaning arrangements which in a roundabout way segregate against race and sexual orientation damaged the Fair Housing Act. In the UK, although the Equality Act-2010 spreads individual parties in a work association, what's more, administration clients in the open segment, it doesn't secure borrowers in a private limited company. Additionally, there is a lacuna in the law. The present expression of the Equality Act-2010 discloses that separation possibly happens when 'an individual' treats another negatively. Even though 'individual' isn't characterized in the Equality Act, an 'important individual' is characterized under area 52 of the Equality Act. 'Important individual' is in mention to a person. In that capacity, mechanized choices of chatbots or robo-consultants won't be gotten below the Equality Act-2010. Security may be that as it may, accessible if an arrangement or choice from an administration office or the controller is unfair. The expectation to segregate is immaterial to the obligation under the Equality Act-2010. The lopsided unfriendly effect of one gathering in aberrant separation could be legitimized by the guard of a balanced method for accomplishing a real point' beneath segment 19 of the Equality Act-2010. Fruitful avocation incorporates cleanliness, wellbeing, and safety, and business needs. It would enthusiastic to perceive how applicable government bodies utilize this safeguard if, later on, their choices are held to be in a roundabout way biased because of the AI calculations. Except if the financial culture and guideline development, predisposition will continue. According to principles six of PRIN 2.1 FCA Handbook by The Financial Conduct Authority-A firm should pay due respect to the benefits of its clients furthermore, handle them reasonably. The rule given by FCA must be paid attention to the expulsion of inclination. Specialists of Account and IT should work with controllers to dispense with inclination.

5.5.1 OVERDEPENDENCE ON MACHINES

Generally, huge amounts of Funds are invested in hardware, software, and technology for AI set up, and the independent automatic working of the machines hence creates a big risk of dependence over machines. Undoubtedly, the machines involved in such industries provide efficiency and save time. But, on the other hand, many well-qualified persons already working in the regular industry have to face unemployment due to the installation of robotics and also Any Machine Error during the working or processing by the AI, which will lead to a huge monetary loss for the business.

For example, ICICI bank has announced a "Robotic Arm" on 28th August 2019 to digitize operations at its currency chests. The arm now helps sort over 6 million notes every working day and 1.8 billion annually. This thing proves to be very useful for the bank but, No one can predict that how long will the machine can retain its accuracy As fears are arising that What if the machine someday stops working or there may be fluctuations in the machine that may result in wrong counting or crushing the notes.

5.5.2 DATA QUALITY

The expectation intensity of a calculation is exceptionally reliant on the nature of the information sustained as information. Indeed, even in quality sources, inclinations can be covered up in the information. The time and exertion required assembling and set up a suitable arrangement of information ought not to be disparaged. For the beginning self-driving car industry, for example, the greater part of the exertion is spent on marking long periods of recordings. This need has prompted the formation of a whole seaward industry for video naming. In the money-related industry, the compromise of the information from front to back is, as of now, tricky, and information referential is regularly tormented with quality issues. Having an information quality program set up is essential to any enormous scale computerized reasoning activity.

5.5.3 BLACK-BOX IMPACT

The aftereffects of shrewd calculations are misty and not certain. They convey measurable certainties, implying that they cannot be right in individual cases. The outcomes could have a concealed predisposition hard to recognize. The diagnosing and rectifying of those calculations is unpredictable.

The way that there is no clarification about why the calculation gave a positive or negative response to a particular inquiry can be exasperating for a broker's sound personality. This is regularly a blocking point for the utilization of AI in exchanging.

5.5.4 NARROW CORE INTEREST

By structure, astute calculations are great at taking care of explicit issues and can't go amiss from what they were intended for. A calculation prepared to distinguish suspicious installments would not have the option to identify some other suspicious action identified with exchanging, for example. Furthermore, calculations are reasonable and need fundamental factors, for example, enthusiastic knowledge and the capacity to contextualize data, in contrast to people. That is the reason banking chatbots regularly frustrate: they are "savvy" however need sympathy.

5.5.5 RESPONSIBILITY

The utilization of insightful machines speaks to a test as far as risk: who/what will be mindful on the off chance that something turns out badly? Money related establishments are hesitant to give machines full independence because their conduct isn't completely predictable. They will, in general, keep a human administrator to approve the machine's choices for basic exercises, for example, discharging/blocking installments or approving exchanges, mostly nullifying the point of utilizing a machine in any case. Current consistency and operational security gauges are very exacting; I foresee that they will release after some time when the innovation develops.

5.5.6 LACK OF APTITUDES AND INFORMATION

An unavoidable topic all through the report, the predetermined number of staff that can utilize AI adequately and the absence of usable information will both moderate the reception and effect of AI. The essential gatherings utilizing AI inside budgetary establishments are centered around research and technique or for very special applications. These two variables are interconnected, as the restricted information will obstruct the number of utilization cases that can be investigated with AI, and the modest number of experts with the range of abilities to use AI will oblige the more extensive use of psychological strategies. Although there are many bottlenecks in the way of implementation of AI in the financial services, there are expectations for a bright future in the field of AI as currently, the technological developments are still at a nascent stage, but in the future, with further innovations and advancements in the field of AI it will be quite possible to economically implement AI systems in financial services and governments would be able to formulate adequate regulations when AI usage become a part of day to day life of people.

KEYWORDS

- **advancements**
- **algorithms**
- **artificial intelligence**
- **financial services**
- **machine learning**
- **smart computers**

REFERENCES

1. Temkin, (2017). *Apply for a Loan in Less Than 15 Minutes.* The Times. https://www.thetimes.co.uk/article/fe8ec664-1b98-11e7-a725-c619aa6571c2 (accessed on 1 December 2020).
2. Moyo, A., (2017). *Absa Banks on Robotics, Artificial Intelligence.* (IT Web) http://www.itweb.co.za/index.php?option=com_content&view=article&id=161393 (accessed on 1 December 2020).
3. Karanasiou, A., & Pinotsis, D., (2017). A study into the layers of automated decision-making: Emergent normative and legal aspects of deep learning. *International Review of Law, Computers and Technology, 31*(2), 170.
4. Alison, L., & George, W. L., (2018). Artificial intelligence and augmented intelligence collaboration: Regaining trust and confidence in the financial sector. *Information and Communications Technology Law.*
5. Agrawal, A., Gans, J., & Goldfarb, A., (2016). *The Simple Economics of Machine Intelligence.* https://hbr.org/2016/11/thesimple-economics-of-machine-intelligence (accessed on 1 December 2020).
6. Datta, A., Tschanta, M., & Dutta, A., (2015). 'Automated experiments on ad privacy settings: A tale of opacity, choice, and discrimination. *Proceedings on Privacy Enhancing Technologies, 1*, 92.
7. Arnold, M., (2016). Market grows for 'regtech,' or AI for regulation. *Financial Times.*
8. Andrew, B., (2018). *The Executive Guide to Artificial Intelligence.* Springer Nature.
9. Sulleyman, A., (2017). Google creates AI that can make its plans and envisage consequences of its actions. London. http://www.independent.co.uk/life-style/gadgets-and-tech/news/google-deepmind-ai-artificial-intelligenceresearch-imagination-augmented-agents-sokoban-a7863061.html. (accessed on 1 December 2020).
10. Banking Technology Vision, (2017). *Accenture Banking Report 14.* https://www.accenture.com/t20170322T205838Z__w__/us-en/_acnmedia/PDF-47/Accenture-Banking-Technology-Vision-2017.pdf#zoom=50 (accessed on 1 December 2020).
11. Buchanan, B., (2005). *A (Very) Brief History of Artificial Intelligence, 26*(4). AI Magazine.
12. Brownlee, J., (2016). *What is Deep Learning?* Machine Learning Mastery.
13. Bauguess, S. W., (2017). *The Role of Big Data, Machine Learning, and AI in Assessing Risks: A Regulatory Perspective.* Speech given to OpRisk North America, New York.
14. Bauguess, S. W., (2016). *Has Big Data Made Us Lazy?* The speech was given to the Midwest Region Meeting-American Accounting Association (AAA), Chicago Illinois.
15. Barclays PLC, Annual Report, (2016). 28.
16. Ben, C., (2017). *Lloyds Bank Announces Locations of 100 Branch Closures and 325 Job Losses.* The Independent. http://www.independent.co.uk/news/business/news/lloyds-bank-100-branches-close-200-jobs-lost-cuts-halifax-bankof-scotland-a7667681.html (accessed on 1 December 2020).
17. Carney, M., (2017b). *The Promise of FinTech-Something New Under the Sun?* Speech at the Deutsche Bundesbank G20 conference on Digitizing finance, financial inclusion, and financial literacy. Wiesbaden Germany.
18. Pollman, E., (2018). The rise of regulatory affairs in innovative startups. In: Gordon, S. D., & Christine, H., (eds.), *The Handbook of Law and Entrepreneurship in the United States.* (Cambridge University Press, 2018).

19. Financial Conduct Authority, (2017). *Financial Conduct Authority Provides an Update on the Regulatory Sandbox.* https://www.fca.org.uk/news/press-releases/financial-conduct-authority-provides-update-regulatory-sandbox (accessed on 8 March 2020). (accessed on 1 December 2020).
20. Financial Conduct Authority, (2017). *Financial Conduct Authority Provides an Update on the Regulatory Sandbox.* https://www.fca.org.uk/news/press-releases/financial-conduct-authority-provides-update-regulatory-sandbox (accessed on 1 December 2020).
21. Scopino, G., (2015). Do automated trading systems dream of manipulating the price of futures contracts? Policing markets for improper trading practices by algorithmic robots [2015]. *FLA. L. REV., 67*, 221.
22. Scopino, G., (2015). Preparing financial regulation for the second machine age: The need for oversight of digital intermediaries in the futures markets. *Columbia Business Law Review*, 339.
23. Government Office for Science, (2015). *Artificial Intelligence: Opportunities and Implications for the Future of Decision Making.* https://www.gov.uk/government/uploads/system/uploads/attachment_data/file/566075/gs-16-19-artificialintelligence-ai-report.pdf (accessed on 1 December 2020).
24. House of Lords Select Committee, (2018). *AI in the UK: Ready, Willing and Able?* https://publications.parliament.uk/pa/ld201719/ldselect/ldai/100/10002.htm (accessed on 1 December 2020).
25. Simons, H., (1996). The paradox of case study. *Cambridge Journal of Education, 26*, 225.
26. Bergen, H., (2016). 'I'd blush if I could': Digital assistants, disembodied cyborgs and the problem of gender. *A Journal of Literary Studies and Linguistics, VI*, 95.
27. Interview with Raza Salim and Jennifer Macdonald, Barclays Bank (Radbroke Hall, Cheshire, 23rd March 2017).
28. Interview with Robert Morris, IBM Watson (Telephone interview, 15th June 2017).
29. Bughin, J., et al., (2017). *Artificial Intelligence the Next Digital Frontier?* (McKinsey Global Institute, 2017) https://www.mckinsey.de/files/170620_studie_ai.pdf (accessed on 1 December 2020).
30. Petrasic, K., Saul, B., Greig, J., & Bornfreund, M., (2017). *Algorithms and Bias: What Lenders Need to Know*. White and Case https://www.whitecase.com/publications/insight/algorithms-and-bias-what-lenders-need-know (accessed on 1 December 2020).
31. Knight, W., (2017b). The financial world wants to open AI's black boxes. *MIT Technology Review*.
32. Warman, M., (2012). *Barclays Pingit: Should You Trust Your Mobile with Your Money?* The Telegraph. http://www.telegraph.co.uk/technology/news/9086152/Barclays-Pingit-should-you-trust-your-mobile-with-your-money.html (accessed on 1 December 2020).
33. Manyika, J., Michael, C., Mehdi, M., Jacques, B., Katy, G., Paul, W., & Martin, D., (2017). *A Future That Works: Automation, Employment, and Productivity*.
34. Scherer, M., (2016). Regulating artificial intelligence systems: Risks, challenges, competencies, and strategies. *Harvard Journal of Law and Technology, 29*(2), 354.
35. Mohri, M., Rostamizadeh, A., & Talwalkar, A., (2012). *Foundations of Machine Learning* (1st edn.). Cambridge, MA: MIT Press.
36. Hosea, M., (2016). *How Brands are Using Artificial Intelligence to Enhance Customer Experience* (p. 1). Marketing Week.
37. Reid, M., (2017). Rethinking the fourth amendment in the age of supercomputers, artificial intelligence, and robots. *West Virginia Law Review, 119*, 100.

38. Crosman, P., (2017). *Can AI Chat with Customers as a Teller-and Should It?* (Vol. 182, No. 11, p. 1). American Banker.
39. Puneet, M., (2019). *Machine Learning Applications Using Python.* Springer Nature.
40. Collinson, P., (2017). *Monzo? It Might Just be the Future of Banking.* The Guardian https://www.theguardian.com/money/blog/2017/sep/09/monzo-future-of-banking-smartphone-tom-blomfield?CMP=Share_iOSApp_Other (accessed on 1 December 2020).
41. Darke, P., Shank, G., & Broadbent, G., (1998). Completing case study research: Combining rigor, relevance, and pragmatism. *Information Systems Journal, 8*, 273.
42. Siroya, S., (2016). *A Smart Loan for People with No Credit History (yet).* (TED Talks). https://www.ted.com/talks/shivani_siroya_a_smart_loan_for_people_with_no_credit_history_yet (accessed on 1 December 2020).
43. Dhar, V., (2015). Should you trust your money to a robot? *Big Data, 3*(2), 55–58. doi: 10.1089/big.2015.28999.vda.
44. Guo, Y., & Liang, C., (2016). Blockchain application and outlook in the banking industry. *Financial Innovation, 2*(24), 12.
45. Campbell, A., (2014). *Artificial Intelligence and the Future of Financial Regulation.* Risk.net. www.risk.net/operational-risk-and-regulation/feature/2374890/artificial-intelligence-and-the-future-of-financial-regulation (accessed on 1 December 2020).
46. Markham, J., & Harty, D., (2008). For whom the bell tolls: The demise of exchange trading floors and the growth of ECNs. *J. CORP. L., 33*, 865.

CHAPTER 6

ARTIFICIAL INTELLIGENCE TECHNIQUES OF FRAUD PREVENTION

SHASHANK AGARWAL

Finance Officer (Head of Finance and Accounts), Kasiga School, Dehradun, Uttarakhand, India, E-mails: shashank0711@icloud.com; financeofficer@kasigaschool.com

ABSTRACT

In the current era of technology, the world at large has been benefitted from the technological progression in almost all the sectors. The advancement of technology, on the one hand, has brought about many boons to society at large, but on the other hand, the fraudsters have found new and faster ways to commit the crime. Hence there is a dire need to introduce new methods to detect and control frauds in the current scenario of technological development. The sector of society that is most affected by these Information Technology frauds is the financial sector. These financial frauds can range from small threats like non-payment for e-commerce orders to major coercions like the exposure of customer's credit or debit card details. When fraud is detected, the organizations end up paying for the heavy losses, and their reputation decays. Hence the fraud detection and prevention methods need to adopt new-age technologies like artificial intelligence (AI) and machine learning (ML) to stop the fraudsters from committing heinous financial crimes.

Once fraud is committed, one can only take corrective actions which involve tracking down the culprit and try to recover the defrauded amount. However, the much better option is to prevent the fraud before it occurs, and this exactly is where the role of AI steps in, which is supported by ML, comes into the picture and provides the best solutions available so far. ML can identify legitimate (genuine) and fraudulent behaviors by comparing them with previously unseen fraud tactics. As far as the role of AI is concerned concerning the use of the models defined by ML, AI uses two models to identify fraudulent practices. One is a supervised model, while the other one is an unsupervised model. Though even after employing ML and AI, certain challenges complicate the fraud detection process, such as changing of fraud patterns by fraudsters, lack of technical knowledge to customers, model interpretations, and time needed to develop and review the already set models. However, ML models and AI increase system efficiency and reduce the instances of fraud occurrence.

6.1 INTRODUCTION

The financial sector, which further drills down to include the banking sector, investment sector, insurance sector, consumer finance, e-commerce, real estate, etc., has been greatly impacted both positively and gravely due to the emergence of technological reforms. Where on the one hand, the financial transactions have become easy, fast, and lightening with technological blessing, on the other hand, they are much prone to risk and fraud as compared to any other sector of society. Nowadays, almost everybody uses net banking, mobile banking, and plastic money facilities. Life has become so easy. With one click of the mouse, you can get the things delivered at your doorstep, or the bills and invoices can be paid electronically. All these reforms could become possible only with the advent of technology. However, this advancement has not only delivered us the benefits but the risks too. Risks of pilferage of money and its diversion by unscrupulous people of society.

Finance, which involves the transactions involving money, has been of great interest to all. Even when the financial sector didn't make much use of technology, the financial frauds were never reduced. They used to take various forms such as pilferage, embezzlement of cash, theft, loot, robbery, and many more of such type. This is evident from the history that every crime had its roots somewhere related to money only. The simple human nature is that greed of man for money never abridges, and to get more of it, few people commit the crime. In the past, committing financial crimes used to involve the physical appearance at the site of the crime, but with the advent of technology, the fraudsters have found new ways whereby they don't need to be physically present at a location, but still, they can steal the money belonging to others and sometimes this is very difficult to trace the culprits.

Once a financial crime has been committed, the culprit can be traced, and corrective measures can be actioned. In simple terminology, this is called fraud detection. It involves all the measures taken to detect the already committed fraud. While fraud prevention does mean to preclude the fraud even before its occurrence, both fraud detection and fraud prevention involve insight into the historical data and feeding it to the financial records to spot the potentially fraudulent behavior in the chain of transactions. Once, during the life of a financial transaction, fraudulent behavior is identified, the whole set of transactions needs to be frozen before it is perpetuated.

The financial sector, worldwide, suffers billions of monetary losses coupled with contrary damage to its reputation and customer's faith due to financial frauds. The "Association of Certified Fraud Examiners" (ACFE) presented a report in 2010 on "Occupational Fraud and Abuse," stating that organizations, globally lose an average of 5% of revenues to fraud each year. Based on the estimated Gross World Product of 2009, this figure interprets to total potential fraud loss of more than $2.9 trillion. PricewaterhouseCoopers's (PwC) Internal Audit 2012 survey reports that "Although antifraud roles vary in business today, top management generally owns the antifraud responsibility, the audit committee oversees antifraud efforts, and internal audit provides a critical line of defense against the threat of fraud by focusing on risk monitoring in addition to fraud prevention and detection."

6.2 TYPES OF FINANCIAL CYBER CRIMES

Usually, the top management owns the responsibility for fraud detection and prevention, but the results are unlikely to achieve without the intervention of technology. The financial crimes which are committed by the fraudsters using technological tools are called cyber crimes. The most common types of cyber crimes committed worldwide are:

1. **Hacking:** It involves illegal intrusion into the computer system without permission, consent, or knowledge of the owner, e.g., virus dissemination, e-mail hack, etc.
2. **Phishing:** It involves a criminally fraudulent activity that aims at acquiring sensitive and personal information such as username, password, and credit card and debit card details, including CVV. For example, creating a fake website replica of an original and trusted website and stealing sensitive information.

3. **Keystroke Logger:** It is a fraudulent activity that logs and records the patterns of key strike without the knowledge of the user. Hence the patterns of username, passwords, and other sensitive information can be stolen.
4. **Vishing:** It is the most commonly used fraudulent activity these days, whereby the bank customers are contacted by e-mail or phone calls and are asked to share their account-related sensitive information. Though many financial institutions are creating awareness to curb such malpractices, the desired results do not seem to be satisfactory.
5. **Spyware:** It is malicious software (malware) which spies on sensitive information without being easily detected.

So, what can be done to curb such malpractices? This is a fact that with efforts, financial crimes can only be reduced but not removed altogether. Even if we succeed in condensing the happening of financial frauds, this will be a reasonable success for us. However, to fight one on one with cybercriminals, the systems in place should be well versed with all the latest technological reforms, and expert systems should be put in place which can read the possibility of happening of the crime even before it is committed. Such expert systems should think ahead in time than cybercriminals.

6.3 ADVENT OF ARTIFICIAL INTELLIGENCE (AI) IN THE FINANCIAL SECTOR

How do financial data experts do this? How do they read or guess the possibility of occurrence of a crime even before it is committed? They use artificial intelligence (AI) and machine learning (ML) techniques for fraud detection and fraud prevention.

Few decades down the lane, we witnessed the advent of a new technology, "AI," which was earlier used in science fiction books and stories, and people used to wonder about the experience of seeing such technology working in the real world. Stephen Hawkings once said, "*It's tempting to dismiss the notion of such highly intelligent machines as mere science fiction.*"

The concept of AI has its roots originated in the 1950s when Alan Turing, an English mathematician and computer scientist, developed a test to identify the intelligence similarities between machines and the human brain. He contended that while interacting with machines, if a human can't identify whether they are interacting with machines or the human on the other side, the machine can be called intelligent. Hence after the concept introduced by Alan Turing was accepted worldwide, the official research on AI began at the same time. To create general AI, numerous AI research programs, methodology, and institutions were set up. However, this research failed to deliver instant positive results; rather, it was during the mid-2000s when AI started to come into existence with innovations and huge breakthroughs such as voice synthesizers, speech recognition, robotics, etc. At the same time, Hollywood also realized the importance of AI, and consequently, few Hollywood movies were produced which were inspired by the concept of AI such as "AI," "I Robot," "The Matrix," "Minority Report" and the famous "Terminator" series to name a few.

AI is not limited to robotics only, as was earlier perceived. Rather, this term can be used for all the technologies which can mimic human intelligence. It can belong to any of these three categories, e.g., (a) ML, (b) natural language processing (NLP), and (c) cognitive computing. The world of finance is rapidly transforming with AI and ML technologies.

Nowadays, AI has caught our nerves to the extent that almost daily, we witness some sort of AI technology. Microsoft's Cortana, Amazon's Alexa, Snapchat, and Instagram filters, DeepMind's AlphaGo, Netflix recommendations, Google Translate, Tesla's self-driving cars are the best-suited examples to it. Most of the AI systems have become so flawless that they have passed the "Turing Test," and it is very difficult to differentiate one's interaction with humans or machines at the same time. Moreover, the public at large worldwide has also accepted AI techniques as less time consuming, easy to operate, easy to understand, and effective.

However, every innovation has its limitations too. This is agreeable that AI will further advance and soon may replace human intelligence, but one of the concerns of AI technology is the automatic generation of malicious data. Since machines and AI work on a set pattern, they are unable to differentiate between good or bad. Elon Reeve Musk, CEO of SpaceX, once stated that "AI can be more dangerous and pernicious than the nuclear weapons and could lead to human extinction if we are not careful." Very recently, in February 2018, "The Malicious Use of AI: Forecasting, Prevention, and Mitigation" report was published by a group of leading AI researchers from various institutions, including Stanford University, University of Oxford, and OpenAI, which details about the potential malicious applications of AI technologies as well as interventions, potential solutions and areas for further research.

Consequently, if a financial institution looks to mitigate the risks of fraud, it can apply the techniques of AI and ML, the best available Computer Science technologies.

6.4 ARTIFICIAL INTELLIGENCE (AI) AND MACHINE LEARNING (ML)

Intelligence is something that came naturally to humans. AI technique aims at using the intelligence gifted to humans without human intervention. This is something very amusing that the human brain looked to develop a computer technology that can make decisions like a human brain and that too, without the use of the human brain itself. Hence, AI enables the computer systems to perform actions like a human brain. However, this is true that as in the current scenario of technological development and increased volume of online financial transactions, this would have been very difficult for humans to check for online anomalies in the financial transactions. Hence, AI aims to imitate a human brain in terms of thinking, working, and functioning. Since this intelligence of computer systems is not natural like a human brain is, it is called AI. It is sometimes wrongly argued that AI is a system; rather, AI is a mere technology implemented in the system. Another misconception about AI is that AI and ML are the same things. AI and ML are two buzzwords nowadays, which often seem to be used interchangeably. However, the many differences between AI and ML can be listed, but broadly speaking, ML is an application of AI which aims to learn the patterns of transactions and detect and separate the fraudulent transactions.

ML is the learning in which machines can learn on their own with or without being specifically programmed. It is an AI application that provides the ability to computer systems to automatically learn and improve with experiences. Hence, ML can be understood as a technique of parsing data, learn from that data, and then apply what they have learned to make a wise and informed decision. While ML is a way to build analytic models, AI is about the use of those models. ML helps the data scientists inefficiently determining which transactions are most likely to be fraudulent while significantly reducing the false positives. AI and ML techniques are enormously effective in fraud detection and prevention as they permit for automatic discovery of patterns across large volumes of online streaming transactions. The assumption behind using the ML for financial fraud detections is that each fraudulent transaction has a specific feature as compared to legitimate transactions. However, the technology is databased and needs more and more data of legitimate transactions to be fed into the system. Whenever there is even a slight variation in the path and patterns of transactions, ML catches the same and flags the trigger for the user. If applied accurately, ML, on the one hand, can differentiate between the legitimate (genuine) and fraudulent behaviors while, on the other hand, it can consistently adapt to the new and formerly unseen fraud attempts. It works on the interpretation of patterns in the data and application of data science for incessantly improving the ability to distinguish between normal behavior and abnormal behavior. Depending on the discovered "fraud-like" patterns, a transaction can be accepted, blocked, or even adopted for a manual review by system analysts. However, in this process, millions of computations are required to be accurately and automatically performed in milliseconds. Nowadays, many business houses use the techniques of ML to give a better experience to their customers, e.g., Amazon gives the matching product choice suggestions to its costumers based on their likings, preferences or last searched similar products. Netflix and Amazon Prime use ML to give suggestions to their viewers on movies or shows that they would like to watch based on their genre search. We can further take the example of almundo.com, a popular travel agent business hub from Latin America, which could reduce frauds, chargeback's, and manual reviews by at least 70%due to the application of ML techniques. If fraud and chargebacks are reduced significantly, it leads to better customer satisfaction, optimization of operational and non-operational costs and hence brings about a significant rise in revenues. Reduced manual reviews don't mean that there is no requirement of the risk managers; rather, ML provides a more powerful tool for them to perform their job more effectively and efficiently.

However, Table 6.1 can give us a broad understanding of the differences between AI and ML.

TABLE 6.1 Differences between Artificial Intelligence and Machine Learning

Artificial Intelligence (AI)	Machine Learning (ML)
Artificial intelligence is a computer system	Machine learning is a subset of artificial intelligence.
Artificial intelligence stands for the ability to acquire and apply knowledge.	Machine learning stands for the acquisition of knowledge or skill.
It works as a smart working computer program.	Machine learning works as accepting and learning from the data.
Artificial intelligence aims at increasing the chances of success rather than accuracy.	Machine learning aims at increasing accuracy without giving attention to the success ratio.
The goal is to simulate natural intelligence to solve complex problems.	The goal is to learn from specific data to maximize performance.
Artificial intelligence is all about decision-making.	Machine learning allows the system to learn new things from data without making decisions.
Artificial intelligence leads to intelligence or wisdom.	Machine learning leads to knowledge and application of knowledge.

AI is often classified into two fundamental groups, i.e., applied or general. Applied AI is more common and involves the systems designed to intelligently trade stocks and shares or to operate a self-driven vehicle that would fall into this category. At the same time, generalized AI has led to the development of ML, which is broadly used to detect frauds and errors in financial transactions.

Another concept that is closely interwoven with AI and ML is Deep Learning, which is a subclass of ML. Technically, Deep Learning is a ML technique with capabilities different and unique as compared to ML.

The main difference between Deep Learning and ML is that though ML models improvise overtime, still they need some sort of guidance to work properly. If the prediction returned by a ML model turns out to be inaccurate, then the manual intervention of the system programmer is required to fix the problem causing inaccurate results, while in the case of Deep Learning, the model resurrects by itself. The automatic car driving system is one of the best examples of deep learning.

This difference can further be understood with an example. Suppose we have an automobile system, and we instruct a ML model that whenever someone says "turn on AC," the Air Conditioning should be put on. Here, the ML model will look for different words uttered by people and will search for the word "turn on AC," and if the words match the learning database, the air conditioning will be put on, but what if someone says, "It's scorching hot outside, reduce the temperature." The user wants the air conditioning to be on, but the sentence does not consist of the word "turn on AC," so the conditioning will not be on. That's how Deep Learning is different from ML. If it were a Deep Learning model, it would turn on the air conditioner as a Deep Learning model learns from its own devised methods of learning and computing.

Having understood the terminologies of AI and ML, now the question arises as to how AI helps the prevention and detection of financial fraud?

Since ML is a technique and subset or subclass of AI, the frauds are caught by the application of ML techniques as a part of AI. Now ML can further use the technologies of Deep Learning as well as one of its sub-types.

6.5 ROLE OF ARTIFICIAL INTELLIGENCE (AI) AND MACHINE LEARNING (ML) IN THE FINANCIAL SECTOR

Since the time of the rise of AI techniques, these have become an indispensable part of almost all the most demanding industries, including the finance sector.

Business owners and executives incessantly explore the new AI techniques to use in finance and other areas, and they potentially get a competitive advantage on the market. Here are some of the most popular examples of using AI in the finance sector:

6.5.1 ARTIFICIAL INTELLIGENCE (AI) AND CREDIT DECISIONS

AI can lead to better informed and data-backed decisions as they provide a more accurate and faster assessment of a potential customer. AI provided Credit scoring is based on more logical, automated, and sophisticated rules as compared to those used in traditional credit scoring methods. It helps lenders to differentiate between various applicants such as high-risk applicants, low-risk applicants, credit-worthy applicants who lack an extensive credit history, etc. Moreover, machines have a feature, overpowering humans that these are not biased while taking decisions. Various mobile apps used by lending institutions use AI and ML algorithms to evaluate loan eligibility, etc.

6.5.2 ARTIFICIAL INTELLIGENCE (AI) AND RISK MANAGEMENT

One of the segments of the financial sector, which is vastly benefitted by the advent of AI, is risk management. Machine's enormous data processing power enables machines to handle a vast amount of data in a short period. At the same time, cognitive computing helps in managing both structured and unstructured data at a far lesser time than a human would take in manual processing. A leasing company in the USA, Crest Financials, employed AI in its risk management methodologies and witnessed an immediate and significant improvement in their risk analysis.

6.5.3 ARTIFICIAL INTELLIGENCE (AI) AND FRAUD PREVENTION

Since the introduction of AI, it has proved to be very successful in detecting and preventing financial fraud. One of the expertises of AI in detecting financial frauds is the prevention of frauds related to Credit Cards, which are growing exponentially in recent years due to an increase in online transactions and the wide use of e-commerce. Fraud detection systems analyze clients' behavior, location, computer IP address, and client's buying habits and subsequently devise a security mechanism when something contradicting the established spending pattern occurs. Banks have started employing AI techniques to prevent yet another major financial crime called "money laundering." Through the AI techniques, machines recognize the apprehensive activities over the web and help to reduce the costs of investigating the alleged money-laundering schemes. One case study reported that due to the use of AI, the investigative workload on money laundering had been reduced by nearly 20%.

6.5.4 ARTIFICIAL INTELLIGENCE (AI) AND TRADING

Machine data-driven investment trading, which is also alternatively termed as quantitative, algorithmic, or high-frequency trading, has expanded rapidly over the past few years across the world's stock markets as AI offers multiple significant benefits in the trading business. Intelligent trading systems are capable of monitoring both the structured and unstructured data in a fraction of seconds. While structured data includes databases, spreadsheets, etc., the unstructured data may include social media, news, etc. It offers faster data processing and faster trading decisions resulting in faster trading transactions. No doubt, the ML and AI-based predictions for the performance of a particular stock are more accurate as the ML algorithms use the past data to test the credibility of the trading systems. In addition, based upon the investor's short-term and long-term goals, AI offers recommendations for the strongest portfolios to invest in. Nowadays, most of the financial institutions trust AI to manage their entire portfolios.

6.5.5 ARTIFICIAL INTELLIGENCE (AI) AND PERSONALIZED BANKING

In the banking sector, the technology of AI has proved to be a boon as it provides comprehensive self-help solutions to its clients. With this advantage offered by AI, the overall workload of the call-centers has reduced significantly. Voice-controlled virtual assistants and bank self-education features are few advancements, which AI has brought to the banking sector. Customers nowadays can check their account balances, schedule payments on a specific date and time, look up account activity, etc., with one tap on the bank app. Almost all the banking apps provide clients with features such as reminders to pay bills, recharges, virtual deposits, plan, and manage their expenses, and interact with the bank in an easier and more streamlined way. Nowadays, just with the advent of AI, the customer's queue in banks has reduced significantly.

6.6 HOW DOES IT WORK?

Arthur Lee Samuel, America-based AI, and computer gaming innovator, stated that ML a field of study which grants computers the capability to learn without explicit programming. In a very common language, ML automates and improvises the computer learning process without the intervention of human experts. The ML process starts with feeding good legitimate data into the system and then training the machines to automatically build the ML models with different algorithms. If there is any variation in the results while processing new data into the system, the ML model can identify the deviation from the originally fed data and results and thus can declare the data as fraudulent. The idea behind this methodology is that fraudulent transactions possess certain characteristics different from good and legitimate transactions. Based on a thousand pieces of information, ML algorithms recognize transaction patterns in the data, which discern the fraudulent transactions from legitimate ones. Whenever a transaction is initiated by a customer, the ML models thoroughly scan his/her profile and search for suspicious patterns. This can better be understood with an example. While students prepare for the examinations, they are not supposed to rote the subjects; instead, they try to develop a complete understanding of them. They feed their brains (machine) with questions and answer from different sources (a good amount of high-quality data) and train the brain to use logic to solve different types of unique questions with a general understanding of the topic. In the process, they solve numerous practice test papers and questions asked in previous examinations. The performance in such papers is compared with the answer keys, and accordingly, the performance is measured. If some more improvement is required, more quantity of practice papers is attempted, and accuracy is achieved over the period. Hence, more confidence is gained with more accurate performance. That's exactly how the ML models are built. Machines are fed with both previously experienced input and output data, and then real-time data is tested by comparing its results with the actual output.

Hence, the ML implementation process consists of the programming of data into the system. The programming of ML models is vastly different from traditional programming in the terms that in Traditional programming, the programmers first feed the data (inputs) and program (logic) and then check the results (output). While in ML models, programmers feed the data (inputs) and results (output), run it on a machine hence enable the machine to create its program (logic).

At present, ML is being applied in nearly all areas of business, e.g.:

- Virtual personal assistants such as Siri, Google Now, Google Home, Amazon Echo, Alexa, etc.;
- GPS traffic predictions using maps;
- Online transportation stimulus such as Ola, Uber giving fare recommendations based on pickup location and destination;
- Face and figure print recognitions;
- Facebook and LinkedIn suggesting known contacts based on the friend list;
- Recommendation of similar products and pins by various online e-commerce websites;
- Video surveillance;
- Filtering of spyware and malware and notifying the user;
- Online and virtual chat option to consumers on the website;
- Image and speech recognition;
- Credit scoring;
- Medical disease diagnosis and many more.

Nowadays, machines can drive a car, pilot, and aircraft, interact with humans like humans such as Siri, and even recognize sentiments. In addition, no doubt, ML technology has also proved to be extremely effective in the fields of fraud detection and even fraud prevention.

Since technological advancement is available to all, online financial frauds have become more urbane as this advancement is available to fraudsters as well. Hence to move one step ahead of fraudsters, companies are required to accurately analyze more and more data to efficaciously sense the fraudulent attempts. The difference between the skills of humans and machines can be understood in the sense that a skilled analyst can analyze, say, up to 50–100 chunks of information, while ML can analyze thousands of information pieces instantly. Moreover, the traditional approach of fraud detection has the demerits that make it less effective such as:

1. Whenever there is a need for a new rule to be implemented, the traditional manual approach will have a time lag between new rule

identification and it's filed implementation. However, machines do these things very instantly.

2. Fraud detection rules are generally created by humans who have their own experience and analytical skills. Nowadays, since online fraud attacks have become more urbane, the rules have also become more and more complex over the period. In the traditional approach, there can be a lack of understanding of the rules due to personal analytical skills, which can lead to an increase in false positives and, ultimately, more money losses.
3. Many expensive "static rules-based systems" seriously depend on human labor. However, in the case of machines, once a program for fraud detection is developed, it doesn't incur high recurring costs in the future.
4. Fraud detection rules grow to uncontrollable sizes depending upon the nature and severity of the problem at hand. It involves a lot of unproductive time when the programmer is left with more rules to analyze. However, with ML in place, the programmer can conclude things more quickly.
5. Every fraud detection strategy is based on predictions of fraud attempts and the history of fraud occurrence. In the traditional approach, this becomes very cumbersome to keep track of various types of fraud attempts and trials. However, ML models devise a strategy based on key performance indicators (KPIs) and already generated fraud attempts predictions. Using the ML models makes it possible to foreknow the refusal and acceptance levels of logic. This enables the programmers to understand how many fraud transactions will be caught at what level of refusals.

6.7 PREDICTION OF FRAUD BY ARTIFICIAL INTELLIGENCE (AI)

The process adopted by every ML process in terms of Fraud prediction starts with the identification of project objectives and ends at fraud detection model upgrading. In between, there are lots of data preparation, data feeding, model building, and prediction stages, which ultimately result in building one such system which can predict and prevent online frauds.

Step 1	**Step 2**	**Step 3**	**Step 4**	**Step 5**
Project objectives identification	Data preparation	Machine model building	Making a prediction	Model upgradation

- **Step 1: Identification of Objectives of the Project:** At first, the business houses need to identify the business objectives to be achieved. Such objectives may vary from organization to organization. However, in terms of online financial fraud detection, these objectives may include, for instance:
  1. Minimizing the false positives;
  2. Keeping the operational costs of manual review at a minimum;
  3. Minimizing the estimated chargeback ratio, etc.

  Once objectives are defined, a few important tasks to be performed would be:

  1. Identification of main project KPIs;
  2. Identification of main sources of revenue and the forces blocking the revenue;
  3. Determining the success criteria of the project;
  4. The one and most important objective would be to predict whether an attempted transaction is genuine or a mere fraud attempt.
- **Step 2: Preparation of Data:** Data preparation involves the gathering of historical data about previous historical fraudulent events. Hence, the more the data is available to analyze, the better it will be. Such data could include the frequency or value of the transactions, chargeback reports, the history of the previous purchases, etc. Once this data is gathered, lining the objectives of the organization, it is cleaned, prepared, and filtered into a form that can be understood by the machine. This is the most time-consuming process of the whole ML process and involves technical expertise as well.

  The outcome indicates whether a particular transaction turns out to be legitimate or fraudulent. ML algorithms look for fraud patterns without paying attention to the class of transactions considered fraudulent by the business houses. The quality of the desired results determines the accuracy of the ML algorithm. In addition, of course, a ML model is considered to be stronger if it can identify more and more categories, e.g., genuine customer, frequent visitor, a regular customer, or a fraudster.
- **Step 3: Building the Machine Learning (ML)Model:** The whole process of ML is about finding what we actually require from the model and then building the model based on expectations. These expectations, no doubt, depend upon the past historical experiences

of the organization. Ideally, once the information about a new transaction is provided to the ML model, the model generates the recommendation on the transaction being legitimate or fraudulent.

During the process of building the ML model, the programmer takes the datasets from Step 2 (Data preparation) to identify what characteristics of data are marked as fraudulent and what are the best fraud forecasters. Analyzing and drawing an evocative conclusion is not a trifling task under the conditions where there might be hundreds of features describing transactions, customers, and their behavior.

However, the process of building the ML model requires proper technology and domain knowledge for data scientists to know how to gather the different kinds of data, identifying the most suitable modeling technique for a particular class of business, the best set of model parameters, etc.

- **Step 4: Making Predictions:** Once the process of building the ML model is complete, the next process is to make it work as per business requirements. This is again a crucial step as if this step fails, the whole process of the ML building process is defeated. The foremost requirement under this step is to deploy and integrate this model with IT infrastructure.

  Every time a customer does a transaction with the company or even searches for a product, the information about this transaction or search is sent to the ML model. The model, upon receiving the information, performs the "data scoring" process and generates a recommendation about approving, blocking, or marking for manual review.

  During a manual review, as stated above, if a suspicious transaction is marked as false positive (legitimate), the ML model uses this information to make a better and more informed decision in the future.
- **Step 5: Upgrading of Models:** The real-time ML Models are designed in such a way that they are continuously reviewed for their performance and are constantly reskilled to be able to spot new emerging fraudulent patterns. This is a fact that ML models have become more effective with more and more data analysis over the period.

  As fraud attacks are getting more sophisticated, there is a need to study more data to enable them to detect the fraud successfully. Only feeding of more data is not sufficient. Some other hardware features such as GPU capabilities of the system, system's processing power, type of connection, use of a VPN connection, etc., can bring about a lot of new acumen about how the consumer will behave. This overall increases the accuracy of prediction.

6.8 SUPERVISED AND UNSUPERVISED MODELS OF MACHINE LEARNING (ML) IN PAYMENT FRAUD DETECTION

Payment fraud is the major area where AI and ML have contributed to their positive results. When a consumer gets a call, text, or an e-mail from their Banks asking them to validate a transaction, they may not even think that this excellent customer service is driven by a brilliant set of ML algorithms.

Unless the data scientists have a proper understanding of fraud-specific data science techniques, the results which come out of the ML process will not hold and accurate. This may further result in a very costly and grave mistake which is difficult to recoup. Following methods can be employed to get the best results of the ML technique and its implementation in a system.

As financial frauds are multi-dimensional these days, so as the ML techniques need to be very proactive and multi-dimensional. Each historical fraud case should be analyzed, and accordingly, an expertly crafted anomaly detection technique should be devised which can address the problem at hand. Hence, both supervised and unsupervised models of ML play an important role in fraud detection and prevention (Figure 6.1):

1. **Supervised Learning Model:** It is the most common form of ML widely used across all disciplines, which are trained on a bigger set of properly "labeled or tagged" transactions. Each transaction is tagged as either bad (fraud) or good (non-fraud). Massive amounts of such tagged transactions are ingested into the system to learn the non-fraudulent patterns. While developing the supervised models, the clean and relevant training data is a pre-requisite. The main drawback of the supervised learning model is that it can't detect fraud if the historical data doesn't contain a specific type of fraud patterns. Hence in this type of ML model, the inputs and possible outputs are ingested into the system, and the system, in turn, relates the new transaction's inputs and attempts to identify similar outputs. Hence these are the predictive models with labeled data. The most common supervised learning algorithms are nearest neighbor, decision trees, linear regression, SVMs, neural networks (NN), etc.
2. **Unsupervised Learning Models:** These are designed to identify anomalous behavior in cases where tagged, or labeled transaction data is either insufficient or is non-existent. In other words, the systems are trained with unlabeled data. In such cases, a form of self-learning by machines is employed to identify patterns of fraudulent or unfraudulent behaviors. Whenever an unseen fraud tactic

is adopted by fraudsters, the unsupervised model comes into the role to detect the same because these models catch the behavioral anomalies by identifying the transactions. There is no one available to train the model in these cases whilst the models learn with its own experiences and learning of historical patterns. Inputs are ingested in unsupervised learning models; however, there are no defined outputs. Hence these are the descriptive models with unlabeled data. The most common unsupervised learning algorithms are clustering algorithms and association rule learning algorithms.

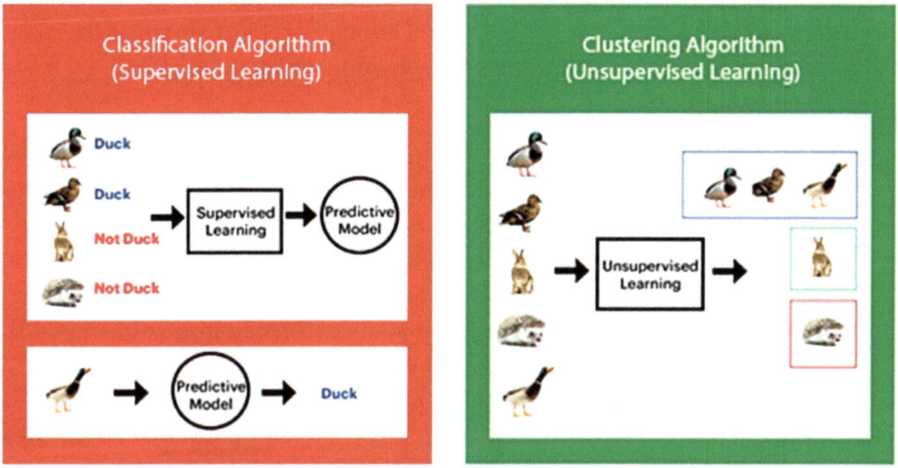

FIGURE 6.1 Supervised and unsupervised models of machine learning.

3. **Semi-Supervised Learning Models:** This falls in between the supervised and unsupervised learning models. When there are certain limitations to use the supervised and unsupervised models, such as cost, the involvement of human experts, the semi-supervised models provide the best solution at affordable costs. These models work on the assumption that even though unlabeled data are insufficient to establish the set pattern but still data carries all relevant information about the group parameters.
4. **Reinforcement Learning:** These models look at using the observations gathered from interaction with the environment and suggesting the best available solutions. In other words, they learn from the experiences of others. These models and algorithms keep learning from the outside environment until they become self sufficiently

learned. Reinforcement learning models are best used in the case of computer applications such as self-driving cars, games of chess and other board games, etc. The most common algorithms of reinforcement learning models are "Q-learning," deep adversarial networks, and temporal differences.

6.9 WHY USE MACHINE LEARNING (ML) IN FRAUD DETECTION?

This is obvious that machines offer better and fast processing of data than human. Applying the cognitive computing technologies (a subset of ML algorithms), the fraud can be detected and even prevented amongst the mass volume of data. Few major factors which support the use of ML in fraud detection can be:

- The efficiency with which machines can perform repetitive tasks. Machines can perform the tasks of data analysis and will render only a few tasks where human insight is required.
- The speed with which machines can devise the complex rules. Building and analyzing the already built rules is a complex and time-consuming task in which machines can perform in microseconds. ML is capable of real-time evaluation of huge transaction data.
- Effectiveness of ML algorithms increases as more and more data is infused into the system. More data infusion guarantees more informed fraud detection. Once machines are trained to identify the genuineness of data, they can filter the new and raw data into fraudulent and legitimate categories.

6.10 CONCLUSION

Even after the advent of AI and its contribution to the financial sector, it is reshaping itself as per the needs of the landscapes of the financial industry. There are high hopes from AI for the increased security measures, especially blockchains and cryptocurrency, expands. As we can see, the benefits offered by AI in the financial service sector are manifold and tough to overlook. According to Forbes, 65% of senior financial management teams expect positive changes from the use of AI in financial services, which has been created by the AI environment. However, by 2018 only one-third of financial institutions started implementing AI in their company's IT processes. Most of the financial giants still fear caution, time, and expense, and the challenges of implementing AI in financial services. However, this is the truth that in the current scenario of technological advancement, a human brain can't perform all the tasks which a machine can do. In addition, the most important task is to gain the confidence of consumers and prevent financial fraud. The fact is that the volume of online transactions and information can only be handled accurately by technology. However, every technology has its flaws also which need to be handled with utmost care and foresightedness. Firstly, the process of building the ML Models involves the involvement of the human brain at initial stages; also, these become very costly over the period. The analysts who perform and analyze the ML algorithms are costly experts, and simultaneously there are recurring system requirements also. However, these system experts create the rules based on their knowledge, experience, and skills, which are bound to be limited. Still, with these limitations, ML is a better solution to fraud prevention as compared to Traditional Learning. The biggest benefit of ML is that the more data ML algorithms process, the more they learn, and the more they become better [1–11].

KEYWORDS

- **artificial intelligence**
- **financial sector**
- **fraud**
- **machine learning**
- **supervised model**
- **unsupervised model**

REFERENCES

1. Roberto I. *Machine Learning vs. AI, Important Differences between Them.* https://medium.com/datadriveninvestor/differences-between-ai-and-machine-learning-and-why-it-matters-1255b182fc6 (accessed on 1 December 2020).
2. Vian, C. *Artificial Intelligence and the Future of Financial Fraud Detection.* https://www.forbes.com/sites/theyec/2018/06/04/artificial-intelligence-and-the-future-of-financial-fraud-detection/#7847550d127a (accessed on 1 December 2020).

3. Horan, T J. *5 Keys to Using AI and Machine Learning in Fraud Detection.* https://www.fico.com/blogs/5-keys-using-ai-and-machine-learning-fraud-detection (accessed on 1 December 2020).
4. Abhinav, J. *How Machine Learning Helps in Fraud Detection.* https://medium.com/ai-techsystems/how-machine-learning-helps-in-fraud-detection-52f29be6d201 (accessed on 1 December 2020).
5. Candice, S. *AI Today-The Growing Impact on Finance Industry.* https://medium.com/shufti-pro/ai-today-the-growing-impact-on-the-finance-industry-ab328fcb6370 (accessed on 1 December 2020).
6. Retail Technology Trends. *Artificial Intelligence and the Future of Financial Fraud Detection.* https://medium.com/@retailtech/artificial-intelligence-and-the-future-of-financial-fraud-detection-b82fa822053e (accessed on 1 December 2020).
7. Mohit, G. *What is Machine Learning.* https://www.geeksforgeeks.org/ml-machine-learning/ (accessed on 1 December 2020).
8. Bernard, M. *What is the Difference between Artificial Intelligence and Machine Learning.* https://www.forbes.com/sites/bernardmarr/2016/12/06/what-is-the-difference-between-artificial-intelligence-and-machine-learning/#3fac531a2742 (accessed on 1 December 2020).
9. Hope Reese. *Difference between Artificial Intelligence and Machine Learning.* https://www.geeksforgeeks.org/difference-between-machine-learning-and-artificial-intelligence/ (accessed on 1 December 2020).
10. Nethone, *A Beginner Guide to Machine Learning in Payment Fraud Detection.* https://medium.com/@Nethone_/a-beginners-guide-to-machine-learning-in-payment-fraud-detection-prevention-360c95a9ca54 (accessed on 1 December 2020).
11. Anthony, Z. *The Rise of Artificial Intelligence.* http://www.immpressmagazine.com/the-rise-of-artificial-intelligence/ (accessed on 1 December 2020).

CHAPTER 7

A CASE STUDY ON DEVELOPING AN END-TO-END SOLUTION TO IDENTIFY LOAN ELIGIBLE CUSTOMERS AND ASSIGNING THEM CREDIT LINES BASED ON THEIR PREDICTED INCOME

GAURAV NAGPAL,[1] ANKITA DHAMIJA,[2] and DIVYANSH GUPTA[1]

[1]*Birla Institute of Technology and Science, Pilani, Rajasthan, India, E-mail: gaurav.nagpal@pilani.bits-pilani.ac.in (G. Nagpal)*

[2]*Lingaya's Lalita Devi Institute of Management and Sciences, New Delhi, India*

ABSTRACT

The objective behind this project is to identify eligible, creditworthy customers to whom the B2B2C lending organization can extend a line amount offer. The project requires three prompt processes that will identify eligible customers. The first process is to develop an underwriting score model to identify the customer's intention of repayment. The second process is to develop an income prediction model to predict income for these customers as it will show whether the customer will be able to repay the loan or not. This predicted income would also help in determining the line amount which should be offered to them. In addition, the last step is to apply the conditions of the policy that have been overlaid.

7.1 INTRODUCTION

In this digital age, customers desire to have more personalized service and hassle-free experience while applying/enquiring about their financial products. Customers also expect these financial services/products to be accessed/available through multiple platforms insecure manner. This requires companies/firms to have a strong foundation for automation and Data Science.

Data Science is an inter-disciplinary domain, which makes use of a variety of scientific methods, processes, predictive modeling, machine learning (ML) algorithms, and prototyping to extract knowledge and insight from structured and or unstructured data.

Loans generate a significant portion of total revenue for any financial firm. Loans are like an investment (like purchasing shares and equity) in a broader sense, so if a bank/NBFC wants to make loans a good risk which generates good returns, a lot of data analysis is required, but in real-world data, availability is often a constraint. This constraint creates hurdles in loan creation, and a lot of people who need loans do not get them.

Data insufficiency is not a constraint in this project. The B2B2C lending organizations have many associated partners through which they provide loans to the customer. For this project, an associated partner has rich transaction data for the transacting customers on their POS terminals. The associated partner in this project wants to utilize this transaction data along with the credit bureau data for effectively and efficiently identifying the customer who needs credit and is capable of repayment. After the identification of the eligible customers, they will be offering a credit line amount with a detailed repayment schedule. The loan gets ultimately disbursed to only those customers who accept terms and conditions and the credit line amount.

This action research project requires three prompt processes that will identify complete offers to eligible customers.

1. The first step involves developing an Underwriting/Application score model that will predict the customer's intention of repayment.
2. The second step involves developing an income prediction model to predict customer ability of repayment and assigning credit line.
3. The third and final step involves applying a group of overlaid policy rules to predict/identify customers who are a good credit risk.

Predictive modeling has been done to accomplish this project. Predictive modeling involves various processes like data ingestion, feature engineering, analysis of data, and the application of various ML algorithms for closer fitting between the observations and the actuals. This project has been very useful to the firm since it came up with a scientific decision generator on providing the line of credit to cardholders (debit/credit).

If enough data is not present, the customers can be under-written by appropriate leveraging of the alternate data. The detailed process is mentioned in the methodology section.

7.1.1 PROBLEM STATEMENT WITH OBJECTIVES

The main incentive for this project is to identify customers who need credit and are capable of repayment. The company is willing to invest in good risk. This incentive is supported by the fact that the associated partner has sufficient transaction data for targeted customers.

The steps mentioned as under were followed to achieve the above-mentioned objective:

- Developing an underwriting/application score model to predict customer's intention of repayment;
- Developing an income forecasting model to forecast the customer's ability to repay and to make a decision on assigning the credit line;
- Overlaid policy rules to predict/identify customers who are good credit risks;
- Out of time sample testing to observe how the model is performing;
- Model recalibration (if required).

7.2 METHODOLOGY

This section will broadly explain the processes which are going to be used. This section is going to be categorized into three parts as mentioned below: formulating a model that assigns application scores or under-writing scores; coming up with an income-forecasting model; followed by applying the terms and the conditions of the policy to arrive at the final number of eligible customers for loan offers.

The data used for this is not only that of the transactions done by the customer but also the credit bureau data. Whenever a customer makes any transaction at any of the partner's POS terminal nationwide, the transaction data gets generated. Credit bureau data has all the bureau data, i.e., customer's data from their previous and ongoing loans, payment history schedule, and many more fields.

The work of this project is mainly done on the two software, i.e., MS-Excel, Python. The main software used to write the model development script in python. Python is an interpreter, a high-level programming language for general-purpose programming. It supports multiple programming paradigms, including object-oriented, imperative, and functional, and many more. Python was considered best for code writing as it is an open-source language with a vast compatible functional library. These libraries provide a wide range of functionality like graphical user interfaces, web frameworks, multimedia, databases, networking, test frameworks, and automation. Some of the important functional libraries are pandas, NumPy, train-test split, and many more. During the development phase, these libraries were extensively used.

7.3 UNDERWRITING SCORE MODEL DEVELOPMENT

7.3.1 WHAT IS UNDERWRITING AND UNDERWRITING SCORE?

This process is about studying a loan application to estimate the quantum of risk involved in issuing the loan. Not only is the credit history of the borrower checked, but it also is a rough estimate of the property made. Underwriting score helps the company in avoiding bad risks or giving credit to customers who are more likely will not be able to pay back the loan.

The process involving creating an underwriting score model is:

- Data sanitization and feature engineering;
- Data formulation/tabulation;
- Data exploration;
- Building the model using the ML.

7.3.1.1 DATA SANITIZATION AND FEATURE ENGINEERING

Both the transaction as well as credit bureau data has inconsistencies present in them. Inconsistencies can be in the form of the presence of numeric values in an alphabetic column, different types of formats like numeric variable represented or have string format, the non-uniform population in several variables, information present in different typography, i.e., lowercase and uppercase which can create duplicates for a single observation (e.g., Mumbai, MUMBAI, Mumbai all will be considered as different values but, they are same) and many more. The following steps will help in overcoming this problem:

1. **Removing Typo Errors and Correcting Typography:** After inspecting the dataset, many variables do have typo errors like the presence of numeric values in an alphabetic column, insertion

of special characters (e.g., &, !, *, %) and presence of lower and uppercase. Therefore, if these types of characters are present, they will increase the no. of values in a variable. These errors are removed from the dataset with the help of codes and other different techniques.

2. **Data Imputation:** This method is used to impute or fill the blank or missing population in the dataset with the use of appropriate value. These values can either be mean, median or anything else, depending on the trend shown in the dataset.
3. **Feature Engineering:** It is a very important step in the development of the model. The data received or tabulated can't be directly used because the dependent and independent variable does not have any correlation. Thus, we can understand how important it can be feature engineering.

It is not only difficult and time-consuming to come up with the features, but it also requires the knowledge and skill of the experts.

The process involves in feature engineering:

- Brainstorming or testing features;
- Determining the features to be created;
- Creating the features decided in the step above;
- Checking the performance of the features created with respect to the model;
- Making improvements to the features wherever required;
- Returning to the brainstorming stage on the creation of new features or editing of existing features.

7.3.1.2 DATA FORMULATION/TABULATION

After the data cleaning and feature engineering on the transaction and credit bureau data is finished, both the transaction and credit bureau data are then merged/joined with the help of a common key/variable present. The dataset obtained thus, contains the details of the transactions as well as credit bureau data. Predictions can be generated using ML.

7.3.1.3 DATA EXPLORATION

Once the data has been formulated, it needs to be explored. This step is almost like a feature-engineering step. The values of a variable from the dataset are plotted on X-axis, and the values of the dependent variable are plotted on Y-axis in a scatter plot. In this way, we are trying to identify which trend, i.e., linear, polynomial, exponential, or power, is showing the best correlation between the independent and dependent variables. If found that a different trend is showing a better correlation than the linear one, a new variable is created.

Figure 7.1 is an example of the type of trends that are generated for dependent (probability of default) and independent variables (a type of financial institution) in the data exploration step. In this trend, there is a very good correlation between the dependent and independent variables. However, all the correlations are not that good.

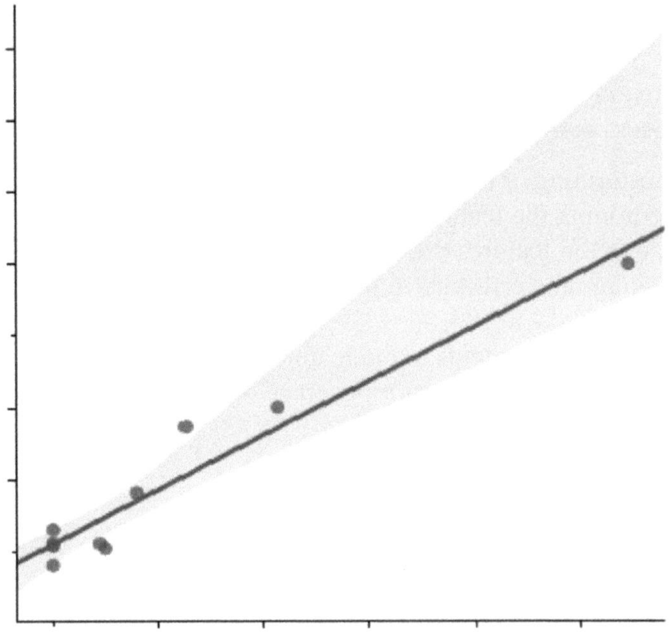

FIGURE 7.1 Type of trends that are generated for dependent and independent variables in the data exploration step.

For example, in Figure 7.2, the linear trend is not the best as there is no direct correlation. Therefore, various non-linear models (for example, power, logarithmic, exponential, and polynomial) are explored, and then the one which shows the best correlation is used in the model development.

The best fit and trend is tried to observe, and the function which can provide is used in the next step, i.e., model development.

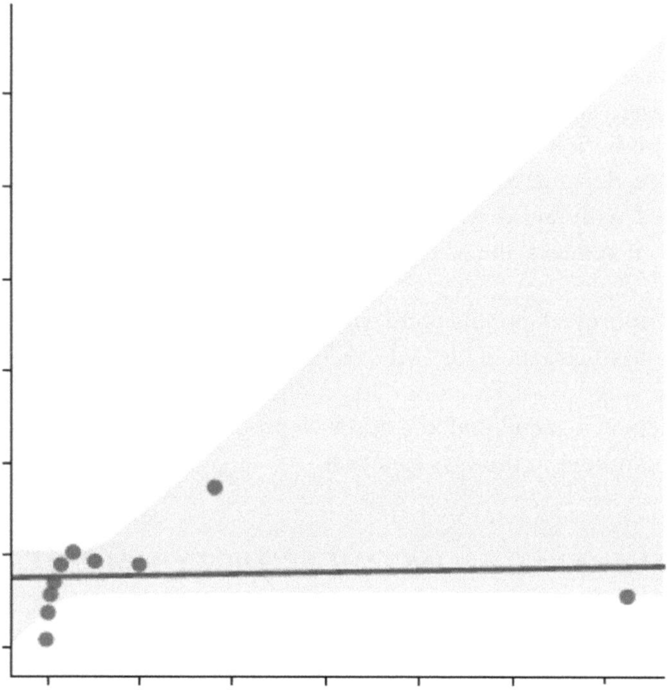

FIGURE 7.2 Correlation in linear vs non-linear trend.

7.3.1.4 MODEL BUILDING WITH MACHINE LEARNING (ML) ALGORITHMS

This is the final step in the model building/development process. In this step, different types of techniques and methods are employed to find the optimal method/algorithm which can provide the best result on a training dataset. ML is a very important tool in model development. It has a wide variety of tests like random forest classifier, gradient boosting method (GBM), Ada boosting method, and so on, which are employed to find the probability of default for the on the test or experimental dataset. For building this model, ML algorithms like random forest classifier, gradient boosting, logistic regression, and ADA boosting were used.

The favorable results were provided by the random forest classifier and GBM. GBM was ultimately selected as it has better discrimination among the variables used from the dataset compare to the random forest method, which is very important because it will make the model exhaustive, and the outcomes cannot change a lot if one or two variables get wrong or incorrect values in the future.

Random Forest classifier is one of the most effective algorithms for predictive analysis. The bagging technique is used by this ensemble algorithm to reduce prediction errors. It determines the statistical relationship between two or more dependent variables, where the independent change variable is associated with and depends on a change independent variable. The way it works is it reduces the variance of the model by way of averaging the prediction.

One of the most popular and widely used methods in ML is gradient boosting classification. This ML technique builds additive models in a forwarding stage-wise fashion. This method does the gradual training of many models in a sequential and additive process. It identifies shortcomings by using a gradient in the loss function.

7.3.2 DEVELOPING AN INCOME PREDICTION MODEL

So, the first process of this project has been completed. The underwriting score model yield the customers who have a good track record in the past for repayment of loans. So, they can be considered as good risks for investment.

Now the question arises that how many lines of credit a customer should be offered. This part is a tricky one because these targeted individuals are not customers of the B2B2C lending organization. Therefore, to assign a line of credit, we need some kind of predicted income.

The forecasted income will assist in the scientific determination of the customer's capability of repayment or tendency of default. According to RBI guidelines, the total EMI that a customer pays cannot exceed 70% of his/her total income. For example, Suppose Mr. X has an income of 50,000, and his existing loan obligation is 20,000 per month. We can't assign a line of credit that has an EMI of more than 15,000 per month.

Therefore, predictive modeling needs to be used, which can constitute credit-bureau data. The income forecast will only be generated for the customers who are approved by the underwriting model developed in the first process.

Developing an income-forecasting model for this dataset is challenging since the predictive modeling requires the training dataset with both the dependent as well as independent variables. Therefore, in this case, the dependent variable is Income, and we do not have that with us. So, it was decided to use the income prediction model, which was earlier developed. The model was developed and deployed successfully and is doing well in predicting income with little standard deviation from the actual one.

7.3.2.1 BRIEF OVERVIEW ABOUT INCOME PREDICTION MODEL

Only the bureau data and the demographic variables have been used to develop the predictive model. Demographic variables are socioeconomic for a customer like Age, Gender, City, State, Contact information, and many more. Bureau variables include debt-related data, inquiry related data, and score data. Model development includes various steps.

All these steps are similar to the ones elaborated while formulating the Underwriting score model. The generalized linear model (GLM) has been used as the machine-learning model. GLM is an effective and efficient tool to solve logistic regression problems. This model is in the form of simple linear regression. However, in our dataset, the dependent variables are not linearly correlated with the independent variable. Therefore, the link function is used to transform the dependent variables so that they can be linearly related to an independent variable. The logit function has been used as the link function in the model building:

- Data preparation/tabulation;
- Feature engineering;
- Data cleaning;
- Data exploration;
- Predictive model building with ML algorithms.

The predictive model was developed only on the bureau and demographic variables. Demographic variables are socioeconomic for a customer like Age, Gender, City, State, Contact information, and many more. Bureau variables include debt-related data, inquiry related data, and score data. Model development includes various steps.

All these steps are explained while developing the Underwriting score model. The ML algorithm used is GLM.

GLM is a very effective tool in solving problems relating to logistic regression. This model is in the form of a simple linear equation with an independent variable, a dependent variable with a coefficient multiplied, and a constant intercept. However, in our dataset, the dependent variables are not linearly correlated with the independent variable, so there is a function called link function which transforms the dependent variables so that they can be linearly related to an independent variable. The link function used in model building is called the logit function.

7.4 RESULTS AND DISCUSSIONS

After completing all the three processes, i.e., formulation of the underwriting score model, income estimation model, and application of the terms overlaid by the policies, we can identify the customers to whom the B2B2C lending organization through its partner can extend an offer of line credit. This line credit amount that will be offered to the customer is within their FOIR limitations. If the customer agrees to accept the line amount with all the terms and conditions like rate of interest, repayment tenure, and so on, he/she will become the customer of the B2B2C lending organization.

This project has huge implications for the future of the company; before this project, the B2B2C lending organizations were involved in Loan Management only; after this, they are now also involved in Loan Origination. In addition, one of the main goals of this project was to find out the customers who are least likely to get the default, so it is expected that the average delinquency of the portfolio will decrease as more and more loans being offered.

KEYWORDS

- **decision science**
- **financial prudence**
- **generalized linear model**
- **gradient boosting method**
- **machine learning**
- **predictive analytics**

REFERENCES

1. *Generalized Linear Model.* https://en.wikipedia.org/wiki/Generalized_linear_model (accessed on 1 December 2020).

2. *Introduction to Generalized Linear Models*. Penn State Eberly College of Science. https://onlinecourses.science.psu.edu/stat504/node/216/ (accessed on 1 December 2020).
3. *Generalized Linear Models*. https://statisticasoftware.wordpress.com/2012/07/16/generalized-linear-models-glz-statistics/ (accessed on 8 March 2020).
4. *Random Forest Classifier from Scikit Learn*. https://scikit-learn.org/stable/modules/generated/sklearn.ensemble.RandomForestClassifier.html (accessed on 1 December 2020).
5. *Random Forest Classifier*. https://turi.com/learn/userguide/supervised-learning/random_forest_classifier.html (accessed on 1 December 2020).
6. *Random Forest*. https://en.wikipedia.org/wiki/Random_forest (accessed on 1 December 2020).
7. *Designing a Restful API with Python and Flask*. https://blog.miguelgrinberg.com/post/designing-a-restful-api-with-python-and-flask (accessed on 1 December 2020).
8. *Building and Documenting Python Rest APIs with Flask and Connexion*. https://realpython.com/flask-connexion-rest-api/ (accessed on 1 December 2020).
9. *Smyth Patrick, Creating Web APIs with Python and Flask*. https://programminghistorian.org/en/lessons/creating-apis-with-python-and-flask#what-flask-does (accessed on 1 December 2020).
10. *Pandas*. https://pandas.pydata.org/ (accessed on 1 December 2020).
11. *Pandas (Software)*. https://en.wikipedia.org/wiki/Pandas_(software) (accessed on 1 December 2020).
12. *NumPy*. https://en.wikipedia.org/wiki/NumPy (accessed on 1 December 2020).
13. *NumPy*. http://www.numpy.org/ (accessed on 1 December 2020).

CHAPTER 8

ARTIFICIAL INTELLIGENCE: A BOON FOR ONLINE BUSINESS

PRIYANKA BHARDWAJ and SHIPRA PATHAK

Chandigarh University, Punjab, India, E-mail: b.piya16@gmail.com (P. Bhardwaj)

ABSTRACT

The application of artificial intelligence (AI) across various industries has revolutionized the interaction of consumers with businesses and vice-versa. However, one sector where its impact has the most potential is in Ecommerce development. In the 21st century, the advancements in AI have not only redefined accessibility from a customer standpoint but in sectors as dynamic as e-commerce; it has changed how businesses analyze their potential and leverage customer data. AI is a futuristic technology that is and will slowly be taking over all types (tiny or giant) e-commerce businesses in the times to come. Most of the companies are already testing these tools on other web pages in one way or the other. Remember, it is time to move with AI for a successful business because you can woo your customers multiple times more effectively and surpass your competitors because the others may be left behind in the race.

8.1 INTRODUCTION

Artificial intelligence (AI), the buzzword of the current century, is the emulation of human intelligence, which contains an unlimited scope and which are smarter than man. The best example of this is 'Sofia,' the first 'robotic citizen,' which has in her a pinch of emotions and her thoughts what we can call extra intelligent.

It is, therefore, not surprising if the e-commerce establishments use AI as a marketing tool to boost their businesses. Similarly, some of the retail industries to use the same to get better results. The AI keenly analyzes the browsing history of clients and then works to suggest to them about similar better options that lure buyers to try out, and this is how the sales are enhanced without much hassles and effort. Let's take some examples of AI used in our daily lives:

- Gmail smart replies;
- Google predictive search;
- Chatbots;
- Google assistant/Siri/Alexa;
- Product recommendations on e-commerce;
- Music recommendations.

Therefore, these were some of the common examples of AI we use every day! Since AI came into the market, all the industries got better with the use of AI.

It is, therefore, not surprising if the e-commerce establishments use AI as a marketing tool to boost their businesses. Similarly, some of the retail industries to use the same to get better results. The AI keenly analyzes the browsing history of clients and then works to suggest to them about similar better options that lure buyers to try out, and this is how the sales are enhanced without much hassles and effort.

When it comes to online music streaming services, the AI plays its role to recommend the users more choices relevant to their preferences, and the music lovers can explore things on a wider platform.

Hence we can derive that AI has gone a long way in revolutionizing the e-commerce business over the globe. In this chapter, we get a glimpse of its role in affecting the retail market.

Overall, AI in e-commerce has brought a great revolution. Let's see how much AI has affected the e-commerce and retail market.

8.2 ECOMMERCE

e-commerce is the form of business that involves buying and selling products and services and making payments for these by the use of internet services (which also refers to e-banking). In other words, it not only covers the physical products but also financial and data transactions are done through the internet.

E-business and e-commerce are two different kinds of transactions wherein business refer to the online sale purchase of goods, commerce include services as well.

8.2.1 *FORMS OF ECOMMERCE*

e-commerce is capable of dealing with various forms that pertain to numerous transactional relationships between businesses and consumers. It can also deal with the products or articles that are a part of these transactions:

1. **Retail:** Anything sold to a customer directly by a business.
2. **Wholesale:** When the products are sold in large quantities to a retail outlet for direct sale to customers.
3. **DropShopping:** When a third party manufactured product is sold to the consumer.
4. **Crowdfunding:** When funds are raised as start-up capital before the manufacture of the product, and the fund is used to launch the product.
5. **Subscription:** Money paid by a customer for an ongoing periodical supply of a product or service for a chosen duration of time.
6. **Physical Products:** It is the physical quantity of product(s) inventory stored for physical shipment to customers as and when orders are placed. This inventory needs to be replenished from time to time to keep the supplies effective.
7. **Digital Products:** These are the tangible downloadable products like templates and courses, or media, which need a license for consumption or usage.
8. **Services:** These are the skill sets or services which are made available to customers in exchange for a fee or sold at a specified rate.

8.2.2 *TYPES OF ECOMMERCE MODELS*

To cover each type of transaction between consumers and businesses, the following e-commerce models are operative:

1. **Business-to-Consumer (B2C):** Buying a product from an online retailer is the best example of this model where goods and services are directly sold to an individual customer.

2. **Business-to-Business (B2B):** Sale and purchase of services like software is the best example of this type of business because it is a deal between one business and another business.
3. **Consumer-to-Consumer (C2C):** OLX or eBay are the best examples of this type where an individual is selling pre-owned goods and services to another individual buyer.
4. **Consumer-to-Business (C2B):** It is the process of selling goods and services by a consumer to a business organization (e.g., an online exposure to a motivational lecture at a cost or a business to use some material for advertisement on a license).

8.3 ROLE OF ARTIFICIAL INTELLIGENCE (AI) IN THE E-COMMERCE INDUSTRY

Ecommerce, also known as electronic commerce or internet commerce, refers to the buying and selling of goods or services using the internet, and the transfer of money and data to execute these transactions. Ecommerce is often used to refer to the sale of physical products online, but it can also describe any kind of commercial transaction that is facilitated through the internet. Whereas e-business refers to all aspects of operating an online business, e-commerce refers specifically to the transaction of goods and services.

The application of AI across various industries has revolutionized the interaction of consumers with businesses and vice-versa. However, one sector where its impact has the most potential is in Ecommerce development. In the 21st century, the advancements in AI have not only redefined accessibility from a customer standpoint but in sectors as dynamic as Ecommerce; it has changed how businesses analyze their potential and leverage customer data.

In this era of technological growth, the advent of AI has paved new business avenues for the e-commerce industry, which has recorded a 72% growth in the recent past. The business leaders are led to believe that the future of e-commerce is even brighter than envisaged because AI has been proved the most essential nutrient for the future business.

AI is a futuristic technology that is and will slowly be taking over all types (tiny or giant) e-commerce businesses in the times to come. Most of the companies are already testing these tools on other web pages in one way or the other. Remember, it is time to move with AI for a successful business because you can woo your customers multiple times more effectively and surpass your competitors because the others may be left behind in the race.

While discussing the benefits derived from AI implementation, we also must feel assured that it is in no way an expensive proposal, as is perceived commonly. The AI is a complicated affair because it is a Robotic device and hence makes life easier after the first phase of installing the system. It may not need much changes in the existing infrastructure, and that is why so many companies are in the test run stage of AI tools within their web pages.

The AI tools based business has registered an increase from 38% in 2016 to 60% in 2017, which is an encouraging trend of this new concept. As a result of this, 77% of companies have confessed a business hike through these improved marketing and sales processes, which is an encouraging statement and promises a bright future for AI systems.

The test runs for AI may even be conducted with an "off the shelf" AI tool, which should be capable of flashing e-mails to various customers through machine learning (ML) recommending them some of the products that could interest them based on their browsing history. This could be done at a nominal cost, and a series of successful tests could pave the way for expanding your AI investment plan to boost your e-commerce business. That will be time to look for different options that could improve customer experience and help you to enhance your trade.

8.3.1 INVESTMENT STARTUP FOR ARTIFICIAL INTELLIGENCE (AI)

It is only a matter of getting started and takes the initiative to implement AI tools in your business. You may not know that some of these tools could already be available to you on your web site if you are hosting the same on Word Press because there are tools that automatically detect spam, delete fake comments, and recommend products or posts to increase customer engagement.

Another place is the Shopify app store, wherein you may find amazing facts about improved e-commerce business through ML implementation. You can either focus on a problem area on your website (like upselling or cross-selling products) or test an AI tool against an existing one that you already have.

The forward-facing AI tools should be followed by the addition of the same at the back end successful implementation of your web site. You must consider testing options, which could result in getting your customers more and more in the sales funnel.

Before you decide on a complete robotic e-commerce business website, you must try different machine tools in phases so that you can pull back or go ahead depending upon the results you derive.

8.3.2 AI FOR E-COMMERCE BRANDS

Having understood the basics of AI and its financial aspects, one can start planning the trial phases in the e-commerce business to attain website improvement goals:

- Recommend customers with items basis frequently bought together or bought as an alternative.
- Install automated chatbots to improve your customer service experience by providing more information to your sales team.
- Send automated engagement e-mails and responses based on customer interactions and information.
- Make use of big data analytics and analysis to know your customer needs and how to market to them.
- Be prompt to respond to comments about your brand online to review strategies.

This just a bird's eye-view of ML and AI-based systems that help businesses to establish a better relationship with their customers. However, a higher-level spectrum has been flashed by Apiumhub about the bigger level AI tools being tried by some companies. With the advent of this technology, e-commerce is certainly moving towards betterment, and as such, a few companies are now on their way to commence with digital marketing or analytics and working toward more complex uses of the same.

8.4 ARTIFICIAL INTELLIGENCE (AI) IN RETAIL STORES

1. **Amazon Go Store:** Amazon, a global level prestigious retail supply chain, has made its name with its efficient services and continuous improvements in working methods by adopting the latest technologies and novel features to their online business house.

   The latest in their list is the "No queue, No checkout" concept, which is identical to a physical store wherein you enter, pick the product, and walk out. They have introduced this concept on their Amazon Go Store, and what all you need is an Amazon account, which permits you entry to this exciting shopping store.

   Here you enjoy the facilities like a credit card system because the product will be billed to your account, and you have the option to pay for convenience. This feature helps busy professionals to save their time as they don't have to go through the check out process by opting for payment modes.

2. **Walmart:** Another online retail chain of similar stature, too have implemented AI features in their store for an automated 'shelf scanning' process which will help them in better inventory management, which is the dire need of the current century supply chain management (SCM) system, to avoid losses due to shelf life expiry and strengthening FIFO where ever needed.

   This efficient Robotic control system will ensure replenishment of items on time; remind managers to price tag the products if the same is found missing or need a change due to price revisions and any other drawback to keep the store in good go condition all the time.

3. **Zara:** It is a world-famous name in the fashion retail business, which is one of the largest stores of its kind. They have implemented an AI-enabled customer help program, which will help the buyer by delivering him his ordered products in a dropbox just on the click of a pick-up code, which is given to them at the time of placing orders.

   In this process, the customer doesn't have to walk between the shelves because the job is done by a Robot. It not only saves time but also avoids mistakes in picking the products manually by the buyers, who may get confused about the lack of identification of the items.

4. **American Eagle:** Here is another example of a similar kind that has been put into practice by an American eagle, the largest ready-made outfit brand.

   In this system, the customer just needs to scan an item from the store, and the store employees will get a notification about customer needs. The outfit will be delivered by the employee right into the dressing room.

   At the same time, the system will recommend more products in similar styles, which could match his choice and expectations. In this way, the customer choice doesn't remain limited to one item as he is getting exposed to more designs available in the store. Certainly, the American eagle is increasing their business through AI and improved customer experience.

8.5 APPLICATIONS OF AI IN E-COMMERCE

1. **CRM:** It's a bygone era when human resources were deployed to engage customers, and they were the bedrock of customer relationship

management (CRM). Disproportionately, big efforts were made to offer superior services to customers. In the present day, AI does the dirty work of sifting through the piles of data and predicting when and what a customer is going to buy.

2. **Internet of Things:** What if everything was synced according to the wishes of the customer. His gadgets, household appliances like washing machines, light air conditioners, and even vehicles. Such a cohesive operation would make life so simple and smooth. This is what the Internet of Things does.
3. **AI in Achieving Sales Goals:** No business is successful until it can sell its products, services, or ideas; e-commerce is all about sales. AI does all the tasks of a typical salesman in the e-commerce market place. It finds prospective customers and even offers after-sales service to them. From micro-level planning to macro planning, AI does it all.
4. **Customer Service:** AI enables the path of success and perfect customer service in this booming economy. Customer satisfaction and responding to their needs is the main focus of AI.
5. **Leverage AI to Enhance Customer Experience:** The most pivotal element for e-commerce companies is enhancing customer experience and creating prolonged engagement. The real world and engaging experiences are missing on Ecommerce platforms. E-commerce portals are much more functional and can offer a much more efficient shopping experience for customers.

   Armed with chatbots, virtual assistants, the businesses have taken customer experience to the next level. Such AI-enabled tools have reduced the need for application interfaces. They, rather have encashed upon the intrinsic potential of hardware like smartphones and speakers, etc.

   The chatbots outdo humans by making the most efficient use of resources at no extra cost. They are also better in data interpretation when compared to humans, as the data outputs are almost error-free. Deep learning and ML empower data analytics to understand underlying user patterns and make the right recommendations for users at the right time.
6. **AI a Value-Added to Sellers Using e-commerce Platforms:** Lack of control over the kind of reviews that will be posted by customers is one big impediment in growth in sales for sellers on e-commerce portals. Fake reviews leave a lasting and negative impact on

customers, and might just dissuade them from buying products. This issue has attracted a lot of attention from e-commerce companies. Amazon, though to a small extent, has been able to address this issue by using AI to place the most relevant and authentic reviews on the top of the list.

Using AI to analyze fake reviews, identify the sellers using unfair means, and selling duplicate products, can be a boon to the e-commerce market places. AI, in such cases, can be effective and efficient at the same time.

AI has been instrumental in bringing greater transparency in business that is carried out in an online marketplace. AI tools like chatbots allow us to scan and analyze millions of products, which is nearly impossible to do with human effort. Alibaba, a Chinese e-commerce platform, has used it to weed out fake products.

7. **AI in Online Retail:** The machines keep a close track of your online activities, like the products you recently viewed; based on that, they pop up sections like recommendations for you or the sections like the customers who viewed this item also viewed.

   Whenever you chat online with any e-commerce portal, a bank, or any other service or goods provider, chances are most likely, and you will be having a chat with the unattended machine. They are intelligent and sophisticated enough to understand your query, no matter how much variance in language exists for a similar query.

8.5.1 PROS OF ARTIFICIAL INTELLIGENCE (AI) IN THE ECOMMERCE SECTOR

1. **Search has Become more Customers Centric Within Online Stores:** The customer drop off-trend has been one of the points of concern for e-commerce websites, which lacked customer-centric search results. To arrest the problem, most websites have introduced AI supported search tools, which helped in improving the customer-centric results. This ML process is capable of providing customers a unique experience whereby the customers buy the products of their choice, style, colors. The products are displayed through images and videos, and hence it helped the customers to shop in a much easier environment. As a result, most businesses are now lured with such effective tools to enhance their sales.

Taking the example of Pinterest, the image recognition software for the chrome extension provides all the help to customers to look for their choicest products across the web with ease.

2. **Customer Experience has Got More Personalized:** Adoption of AI helps in analyzing data, which is not just big but also very comprehensive in terms of choices like brand, style, and colors, etc. This, in turn, provides a personalized experience to customers and helps them by recommending products that might suit their choice.

   Boston Consulting Group (BCG) conducted a comprehensive study on customer behavior and choices and established the fact that those retailers, who resorted to AI tools, have been able to boost up their sales by 6 to 10%, which is a promising sign for the future of e-commerce.

3. **Much Better Sales Processes:** AI, ML, and digital platforms are only a shift from the 'yellow pages' concept of yesteryears, which was the sales-boosting method of the earlier times. While CRM + AI integrated systems are the tools of the sales team to scientifically analyze customer choices, AI is helpful to customers for answering their queries regarding new designs and styles, and it also helps to solve their problems and provide information on new products for sale in the market.

   Getty Images is one such tool that made use of AI to explore the market base for its innovation by scrutinizing data and to identify companies who make use of images in their business development. In this way, the sales team of Getty Images discovered new customers, a two-pronged approach to boost two businesses at the same time.

4. **Targeting Potential Customers:** A huge customer, though it sounds interesting, may have so many challenges too for the business. It is a strong base for the sale of company products and services but may pose a challenge to single out prospective customers because the number is too high to analyze and understand their attitudes.

   A study conducted by Conversica finds that the majority of companies are not following up on inbound sales leads. To resolve this problem, now more and more e-commerce companies have resorted to AI as a tool to accurately analyzing consumer attitudes that are based on their in-store behavior, for which they use facial recognition software and online approach through different offers.

5. **Better and Efficient Logistics:** Some of the very prominent e-commerce companies like Amazon, Alibaba, eBay, and others are

now making use of AI, ML, and robotics for an efficient and seamless delivery process as well as automated warehouses to attain proficiency in speed and efficiency that would cut their costs.

8.5.2 CONS OF AI IN ECOMMERCE

As is human psychology, interacting with machines may not always be a good idea. Chatbots, for example, may look boring to the callers, especially those who have come of age. These people feel uncomfortable with computerized instructions like press 1, 2, or 3, etc. While it is understood that chatbots are an integral part of digital platforms, it would be preferable to have a combination of machine and man wherever possible (Business Encyclopedia).

You can never imagine a world without humans. The science fiction we read or watch presents such a scenario only for thrill's sake because no technology can take away all human jobs. In addition, the most heartening part of it is the marketing job, which can be best performed by humans alone because it is a skill where one needs to change the mindsets of others or take creative decisions, which are the outcome of an imaginative mind. Hence AI can't make up for human creativity and cultural references; it may only offer solutions to certain issues and enhances working efficiencies, whereas humans remain at the very epicenter of the marketing world.

8.6 AI AND ETHICAL CONCERNS

There are certain ethical issues of concern that are the product of enlarging the field of AI and hence need to be addressed:

- Automation and technology may eat human jobs.
- The employees might have to be retrained to enable them to retain their job or at times redeployed, depending upon the kind of technology.
- Fair distribution of wealth created by machines.
- Constant interaction with machines may have effects on human behavior.
- The need to address algorithmic bias originating from human bias in the data.
- The autonomous weapons provided for the security of AI can also cause damage.

- With the future concept of self-learning, smart machines need to palliate so caused consequences.

8.7 5 USE-CASES OF AI IN ECOMMERCE

8.7.1 BETTER SEARCH RESULTS

Many customers drop their purchasing in between for the reason of irrelevant results displayed for a product. To address these crucial issues, organizations are working in the direction of being customer-centric through AI applications so that relevant search results are displayed to them. Currently, e-commerce websites are in the process of developing software that can leverage image recognition and natural language processing (NLP) to produce the desired results more accurately. A very popular search engine has been designed by the name Yandex that implement NLP and Deep Learning to process data of previous searches to identify future searches. The success story of this process can be estimated by the increase in click-through rate, which is almost a 10% hike. Similarly, Clarifai is also working in the same direction to conceive smarter applications that can help to improve the e-commerce business from its present levels. They term it as 'AI with a Vision' that can perceive the future shape of the global business enabling developers to bring in more intelligent apps that should be more customer-centric and hence a base for enhancement of e-commerce business the world over [1].

A demonstration of how Pinterest Lens works. Ecommerce stores are in the process of collaborating with Pinterest for introducing their new product Pinterest Lens that can directly pick up images from their shelves and load them on Pinterest. This will also improve the customer drop out of the problem because it has been observed that they tend to abandon their search for want of desired products. These technologies will provide a two-way benefit by generating more business and easing out customers by providing relevant images on a click.

8.7.2 SHOPPING EXPERIENCE LEVEL 1001

The most effective way to enhance customer experience is to 'make it feel real.' We do it through the information collected by the devices used by the customer, and this is the right data to make the experience feel real. It will lead you to an extremely exciting shopping spree. If you want to verify what is said here, just check your Google Maps Timeline, and you would find all your movements recorded there with dates, etc., [2].

Deep Learning and ML are the technologies that can calculate probabilities of your buying a particular product through data analysis. These are so sensitive that even a casual click on a product could be deciphered.

Boomtrain is another AI search engine, which automatically gets connected to the existing channel used by customers and provides businesses an analysis of customer interaction online and thus provides him relevant images for the products of his choice vis-à-vis exciting and enticing offers alongside dynamic content about the product, its availability, and other required information. It is also designed to accumulate a unified view across different devices and thus analyzes performances on variant platforms. Not ending here, Criteo is another company that helps businesses to prompt personalized online views of the products and related ads. To the customers who had ever visited this website even for a glance. This is how cross-device advertising comes into force to engage shoppers whenever they are online and overall devices, including mobile phones.

AI is a wonderful tool to understand the intricacies of human behavior through deep analyzing insights that are empowered by the fact that AI is capable of scanning data to the terabyte level. Because of this trait, AI can deliver the desired personalized shopping experiences for the customers.

8.7.3 CURBING FAKE REVIEWS

As a part of unethical practices, fake views are uploaded on websites about certain products that compete with other sellers. This happens to ruin the image of the product. It is an unhealthy practice because it may cost companies billions of dollars by way of reduced sales and or in the process of rebuilding the image to win customers' faith.

Although product rating is an integral part of online shopping, yet it is not very true that most customers' are misled by negative remarks about a product. 87% of buyers decide the purchase absolutely with their minds. They just skim through the details once they have decided to buy. Of late, it has been felt that shoppers don't give much weightage to such remarks and go by the gut feel. Currently, AI is also being used to present an analysis of these ratings for buyers' help [3].

Yelp has introduced sentiment analysis technique which involves: Organizing information into data sets relating to business id; Day (date) of posting the review; Star rating (1 to 5); Text and lastly to review the text of the buyers to establish sentimental behavior behind this process.

Facebook also provided an AI solution, 'FastText,' that quickly classifies messages and, in turn, provides an accurate direction for the creation of supervised as well as unsupervised learning algorithms.

8.7.4 SALES FORECASTING

With the change in times from sans technology to high tech devices, AI has assumed the role of God or Charles Xavier to read the minds of shoppers irrespective of their location over the globe. The high cost of marketing expenditure, which was due to speculative sales, has now been optimized as the AI will foretell the sales figures for a product for a given period, thus helping the retailer to store only the desired quantity and void blockage of money on non-selling products. This will also save the cost of storage, damages, and pilferages. Such analytical information will best suit traders dealing with perishable products, which become null and void after the date is over.

Sounds too good to be true? With the advantage of accumulating historical data, AI is the best solution to save physical work and helping the sales force to arrive at correct and more precise decisions. Since the implementation of AI is not very expensive in the present scenario, even small budget companies can afford it. A study of AI aided businesses has revealed the following conclusions:

- Identify fast-selling popular products;
- Suggesting products that should be promoted on a particular date;
- Predicting customers' likely choices;
- Determining a workable MRP for your product;
- Eradicates possibilities of fraud to minimum possible level;
- Providing impetus to SCM;
- Business intelligence is sufficiently boosted;
- Sales would rise, and so makes the profits.

8.7.5 CHATBOTS TO THE RESCUE

In a vast group of 7 billion traders, those e-commerce websites have gained distinction that has added special features to their websites. One of such features is chatbots, as it enhances customer service and satisfaction in one go. The chatbot will ask you a question about the product you are exactly looking for, and subsequently basis your answers, it provides you the near-perfect images and information for you to decide the purchase finally. Let us list some benefits of deploying chatbots:

- Chatbots have helped to build a strong customer base by taking in its fold, even the lazy buyers.
- We have come a long from customary chatbots to intelligent beings capable of understanding and tackling issues that they were earlier incapable of.
- Since a larger population of online buyers (83%) needs real-time support, and chatbots make it possible to provide the same.
- The interactive Chatbots that are empowered with more personalized information about consumers have been able to upsurge sales by 40%.
- Since chatbots are an efficient means of collecting information and feedback, it helps a lot in providing the purchasing patterns of consumers and hence helps in businesses.
- Chatbots are efficient and affordable, and this automated process is active 24×7, saving the huge cost of manpower to answer customer queries.
- Chatbots are gaining ground for the good reason of staying ahead in business as these provide a good marketing campaign.

8.8 FUTURE OF ECOMMERCE

AI is no longer a dreaded word projecting Robots and supercomputers. It is now a user-friendly technology that provides conversation facility with customers without involving humans. Adopting such technology like ML helps both traders and customers to do easy selling and buying. AI is versatile enough to target a large customer base providing them with ads and features of products they are looking for. It is a platform that helps in enhancing business opportunities for online websites.

AI triggers automated e-mails to customers when they abandon their cart to understand their requirements and help them to complete the purchasing [4].

Since the e-commerce business is not destined to look back, AI holds enough potential for deciphering customer behavior and then provides solutions to his needs.

Yes, these devices are the need of the time to keep the ball rolling because AI and ML are designed to create effective e-commerce marketing strategies that drive sales.

KEYWORDS

- **artificial intelligence**
- **chatbots**
- **customer relationship management (CRM)**
- **customer satisfaction**
- **e-commerce**
- **technology**

REFERENCES

1. Mayank, (2019). *Use-Cases of AI in e-commerce*. Retrieved from: https://engineerbabu.com/blog/top-5-use-cases-of-ai-in-Ecommerce/ (accessed on 1 December 2020).
2. Sharma, R., (2019). *Role of Artificial Intelligence in Ecommerce.* Retrieved from: https://hackernoon.com/role-of-artificial-intelligence-in-e-commerce-404cad38e1a (accessed on 1 December 2020).
3. Makhija, R., (2019). *Artificial Intelligence in e-commerce*. Retrieved from: https://www.gurutechnolabs.com/artificial-intelligence-in-Ecommerce/ (accessed on 1 December 2020).
4. Atomic Reach, (2019). *Ways AI Has Changed Ecommerce*. Retrieved from: https://www.searchenginejournal.com/atomic-reach-ai-changed-Ecommerce/310143/ (accessed on 1 December 2020).
5. Business Encyclopedia. *Examples of Ecommerce.* Retrieved from: https://www.shopify.com/encyclopedia/what-is-ecommerce (accessed on 1 December 2020).
6. Expert Commentator, (2018). *AI is Revolutionizing e-commerce Even for Small Businesses*. Retrieved from: https://www.smartinsights.com/Ecommerce/Ecommerce-strategy/ai-revolutionizing-Ecommerce/ (accessed on 1 December 2020).

CHAPTER 9

THE DRIVING FORCES OF FACEBOOK SOCIAL COMMERCE

DEEPSHIKHA SETH and SHALINI SRIVASTAV

Amity University, Uttar Pradesh, India, E-mail: ssrivastav@gn.amity.edu (S. Srivastav)

ABSTRACT

It is seen that trust is a very important factor in someone's life and also in social commerce, which is mainly dependent on the content which is generated by the user. This study focuses on understanding the relationship between social platforms and the intentions of the customer and also will look for a process to discuss their relationship. Therefore, a new idea is produced, which consists of majorly three steps: societal activity, a gathering of informative behavior, and similarity in the same group. These three elements are majorly to build trust and impact all the intentions of the customers from various platforms of social commerce. A major questionnaire is collected from various people on Facebook. The details of the findings show that the belief in all the sites related to the social network has a positive impact on the purchasing powers of customers.

This trust also impacts information, which also strengthens the intentions to purchase. Social impact is also found to create trust and purchase intentions. As per the details, various implications are important that can also create site designs to impact customers. It also tells some deeds for office managers and market services who are involved in actual sales on various online networking platforms.

9.1 INTRODUCTION

Shopping is generally considered a social activity. One would often go with family or friends, get advice, share opinions, and ask details from the salesperson. However, online shopping is different. There is little or no human interaction and social contact. This is one factor that is slowing down the growth of e-commerce to a large extent.

According to Ref. [25], this also results in greater chances of dissatisfaction on sides, higher risks, and the lesser idea of social acceptance-in all, an unfavorable environment for business. As per the start of Web 2.0 technologies, these issues have been addressed to some extent. Web 2.0 technologies allow customers to interact, giving rise to the idea of social commerce. Customers can now gather required information by content sharing and user-generated content, which are the core features of Web 2.0 technologies. This has resulted in positive trends in customer interactions and business results. Online transactions have become more reliable, which in turn has improved the reputation of businesses. The new technology has also made it easier for businesses to contact customers- and vice versa-providing a conducive business environment, especially for new products and brands. The use of Web 2.0 technologies has also proved beneficial to the customers because now they can easily get product reviews and feedback from their peers on the social network. Earlier, customers had to depend only on the one-sided information provided by the sellers on their websites.

The concept and development of social commerce are quite recent, but it has grown rapidly in this short period. This can be attributed to the widespread use and popularity of social media, due to which social communities have developed, and interaction among friends and acquaintances has increased. Consumers use these networks not only to share information about their new buys but also to show their feedback and review ratings. This, in a way, has given a boost to online shopping because people find the opinion of their peers more reliable than that of sellers or market experts they do not know personally.

Businesses, therefore, are not holding back from investing in social networks and social commerce, which have gen them new opportunities to promote their products and services. Several service providers (such as ICICI Bank, Vodafone, and Policy Bazar) and product sellers (such as Samsung, LG, and Unilever) have used social commerce to achieve an upwards trend in their sales figures. Yet, many companies have failed in this area; most of those who were unsuccessful had made the same mistake: they underestimated the need to protect the privacy of the users and ensure security in exchanging information. As a result, consumers could not trust their platforms for making online transactions. Trust is a major reason that leads customers to shop online. Another factor that drives customers to make online transactions is the rampant use of social networking sites (SNSes) by both sellers and people. Sellers use these sites to introduce their products to consumers, but what initiates their interest in buying from an online seller is the content provided by their peers. So, trust in the sellers' message, trust in their security system, and trust in the peers' feedback work together to promote social commerce [20]. However, this theory of trust is opposed by a study which [14] contends that there is not enough research to prove how trust can be attached to online transactions or even intentions to make online purchases. This chapter attempts to rationalize the role of trust and a social networking site (Facebook) in driving customer decisions regarding online purchases and also develop a process through which trust can be used to make an impact on e-commerce.

According to an Indian Brand Equity Foundation report [2], there are 665.31 million Internet users in India (as of June 2019), which is about 40% of the population. The growth of the e-commerce industry in the country has been phenomenal; it was declared as the fastest growing industry in India. From $ 800 million in 2011, the size of the e-commerce industry has grown to $24 billion in 2017 and $ 38.5 billion in 2018. According to the United Nations Industrial Development Organization report, only 9% of Indian Internet users made online purchases in 2013. This number grew almost three times in 5 years, and in 2018, as many as 25% of Internet users in the country were making digital transactions-buying products, paying bills, purchasing tickets, or making money transfers (Figure 9.1).

The growing reach of the Internet and the availability of smartphones can be one reason for this, but it also showed that consumers were no longer shying away from online purchases and making online payments.

There have been several studies on the growth of the Internet in India, how consumers adopted this new technology, and how SNSes are growing [16]. Researchers have explored the difference in trends based on states, cities, age, gender, education, income, and other demographics. Yet, there has been a limited study to examine the factors responsible for the rise in e-commerce, and much less on the impact of SNSes on this industry, which is estimated to witness a steep rise in the next decade.

This study plans to examine this unexplored functionality of social networking and find if it has any role in making e-commerce acceptable and popular. The research was focused on the precise question: How does a social networking site (Facebook) influence users in India to make online purchases? Facebook was considered an appropriate social networking platform for this study because of its widespread use and high popularity among all sections of consumers and sellers, and also because of its involvement in social commerce. This study will provide useful information and implications to the Indian marketers who plan to target Facebook users and turn them into e-buyers.

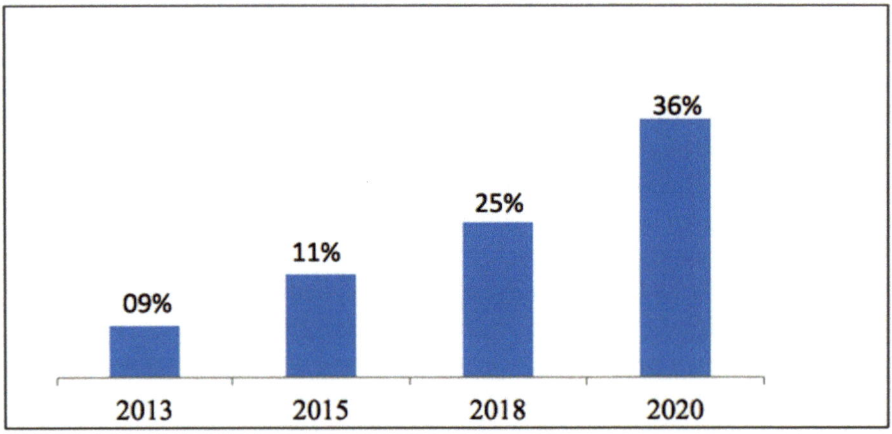

United Nations Industrial Development Organization, 2017.

FIGURE 9.1 Percentage of internet users involved in online shopping.

9.2 SOCIAL COMMERCE

People around the world today are using social media technologies, and India is no exception. The Internet penetration in India in 2019 is close to 450 million (35%); it is increasing by the day and is estimated to reach over 45% by 2021. Due to this growing popularity, many marketers and manufacturers see the high potential here. It increases the visibility of the products, and ultimately their sales. This has added new functionality to social media. Though marketers were using social networking platforms for a long time, of late, they have adopted social commerce as well. In simple terms, social commerce can be described as a combination of e-business and online platforms for social interaction. Some of the experts have defined social commerce as using social media platforms for business or commercial operations [25]. However, another group describes it as the use of Web-based platforms where people interact for selling of goods through online communities and marketplaces. It can also be defined as a segment of e-commerce that makes use of Web 2.0 technologies to help online users interact for selling products and services [23]. Another definition describes it as a division of e-commerce that is aided by both types of social activity or social environments [3, 15]. Social commerce is an extension of e-business that uses social media technologies to provide a communication forum to buyers where they can share their positive and negative feedback about a product to help potential buyers in their buying decision.

To make optimum use of this concept, several e-sellers, such as Amazon and eBay, have incorporated social media apps into their websites to enhance customer communication at the time of making a purchase. For this reason, [23] social and economic features are marked as the two main characteristics of social commerce, saying that it constitutes commercial activities promoted by community interactions and social technologies [41]. Social commerce is that part of e-commerce wherein commercial transactions are supported by social networking.

Online business processes are always the first choice of people. Customers generally go to the market with friends and family and are also helped by shop assistants. The suggestions of others have been a major factor in influencing the buying decisions of customers. However, online shopping is different. It lacks the warmth and sociability of traditional face-to-face shopping. Many find online shopping impersonal, where one is unaccompanied, staring at a screen—which is precisely the reason that makes it spontaneous, something that does not involve much thinking [29] mentions that only the essential features of credit card security and return guarantee are not enough to attract consumers; customers do not favor sites where review comments of past purchasers are missing. Today's customers seek two-way communication between the system and users, where the system can provide the information they want.

Interactions between buyers and sellers in the past have not been easy and involved high costs. Social technologies have somewhat bridged this gap as social platforms help businesses hold regulated communication with buyers without incurring high expenses. Social technologies add an important feature to e-commerce, which earlier had the only objective of making sales. The new technologies, in a way, make shopping a social experience again by helping customers get feedback from earlier buyers and stimulating purchases. Therefore, social commerce helps buyers get a better understanding of a product before making an online purchase.

Huang and Benyouce hint at two main categories of social commerce. One, where Web 2.0 technologies support e-commerce websites, enhance interaction with customers, and help them upload their comments. Here, interaction among the customer is restricted-they cannot comment on other customers' comments or interact with them further by sending private messages. In the second category, e-commerce features are added to the Web 2.0 platform, such as creating a page of a product or company on a social platform like Facebook, Twitter, and Instagram. This is the focus area of the current study. Such pages allow the business to customer and customer-to-customer connection.

Online sellers can also use these pages to update customers about their products by posting news, promotions, images, and videos. These pages let customers add their comments to these posts and share them with friends, thereby helping in marketing the products and influencing their peer group.

9.3 TRUST

Trust is a complicated term and assumes different meanings for different people [25]. The dictionary defines trust as a belief that someone or something is reliable and will not harm you. Trust is the willingness of someone to open to or share information with another person with the expectation that the latter will act in a particular way even when no one is monitoring. Studies have suggested four types of trust: believing various activities, a belief based on the educational institution, change of belief, and people's intention to believe. In the business context, these can be categorized into integrity, ability, and benevolence-trust is how the buyer perceives the seller concerning these three concepts. The integrity of the sellers is their capability to follow buyer acceptable principles and disciplines. Their ability includes product characteristics, competence, and sellers' skills. Benevolence means the readiness of the seller to help the buyer. This study projects the trusting beliefs of one in the promise of another and that the latter, irrespective of the circumstances, will uphold the promise.

Trust is very important in business, especially in the case of electronic transactions where neither the actual product nor the seller is visible. There are also very few regulatory systems for online purchases, and the risk of uncertainty is far greater when compared with a traditional business. That is why trust has been the focus of many types of research in the area of online business and social commerce. Any e-trader looking for long-term relationships must invest in building trust. When trust is high, uncertainty gets low, and this inspires buyers to make an online purchase without hesitation. It is this trust that controls any kind of opportunistic conduct even in the absence of stringent regulations in case of online buying.

Several transaction issues have been reporting in the case of online purchases. These reports impact customers' decision to buy products online.

The buyer and seller are at different locations, there is no direct contact with the seller, and the product cannot be seen. There are questions about the security of the payment methods, and at times incorrect, damaged, or faulty products have been delivered. All of this deters online buyers. This where trust can salvage the situation-trust in the seller's ability, integrity, and benevolence.

Several studies are targeting the value of belief in influencing customers to go ahead with their decisions to buy goods make online [30]. Similarly, there have been several types of research that suggested techniques for building trust. While these studies centered on the idea of using functional features (suitability, ease of use, and benefits) and established quality attributes (standardization, guarantee, and worthiness), the social aspects in online buying were completely ignored. Interactions between buyers and between buyers and sellers are important for building trust in the mind of customers. Social interaction is one way of achieving this, yet, there is little literature on this important part of e-commerce.

Social commerce is a recent phenomenon but has grown quite fast. The social commerce market of the world is estimated to expand at a compounded growth rate annually and is going to reach around 34% by the year 2021, as per the statistics [38]. One of the reasons is an increased rate of social media advertising, driven by the continuous growth of social media users. Though the number of users is increasing, the advertising is increasing, yet many businesses are not holding back from investing in trust-building activities. The benefit of user-generated content can almost turn into a bane in increasing uncertainty when the amount is massive [8]. They have reported that online buying can improve only through a helpful experience between the main parties (buyer and seller). Kim and Park studied the main requirements of belief and its impact on attitude to go for online purchase. They have listed seven factors that can help in building trust-communication, feedback reports, economic feasibility, referrals, and quality of information, structure, and safe business transaction. Customer reviews posted on online platforms also impact potential customers' decisions. Studies have been conducted to show how the relationship between trust and online buying can be measured.

9.4 STUDY MODEL

This study proposes a study model (Figure 9.2) that can be used to describe links between trust and online purchase decisions through SNS. This model uses three main concepts:

1. **Information Seeking:** This highlights the need for collecting information from various channels in an SNS.
2. **Social Presence:** These points to the social interactions on an SNS.
3. **Familiarity:** This includes awareness of the various features of SNS that can be used profitably.

With SNSes, sellers have multiple opportunities to share product information with the buyers [13]. Customers can acquire information through various forums, review sites, comments, ratings, recommendations, and communities. What makes peer-generated content on a forum reliable is the familiarity of the SNS, which can influence purchase intentions positively [38]. The seller's social presence in itself can motivate customers as it gives them confidence that the seller can be contacted easily in case of an issue.

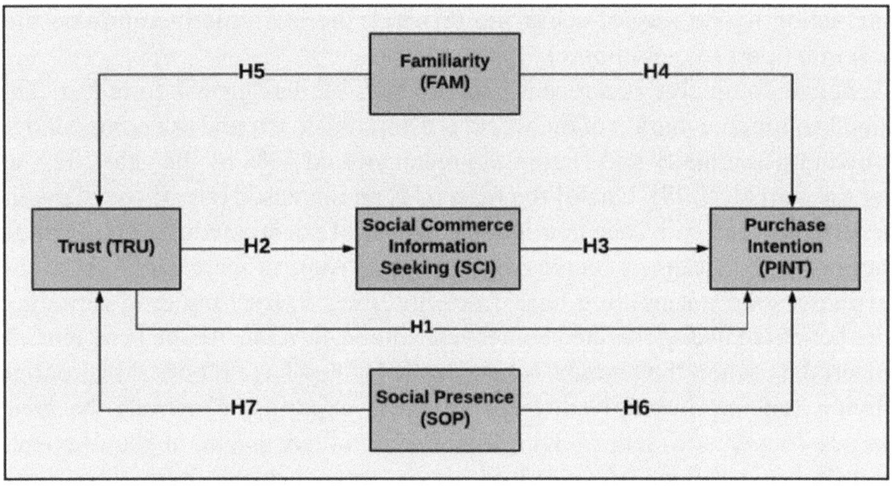

FIGURE 9.2 The study model.

9.4.1 ONLINE CUSTOMER PURCHASE IS INCREASING DUE TO BENEFITS RECEIVED BY SNS

The intention to purchase is the customers' willingness to buy a product [1]. It is said that buying behavior can be judged from behavioral intentions. The buying behavior of customers is based on their intention to make a purchase. As there is a lack of face to face interaction between seller and buyer in the case of online transactions, it leads to social complexity and uncertainty and hints at the unpredictable and opportunistic behavior of the seller. According to Ref. [22], the buyer can trust an online seller if the former perceives that the seller can provide the product with the desired or better features. Trust, therefore, is the prime factor that impacts buyer's buying behavior, which ultimately translates into purchase intentions. Trust also decreases the vulnerability of customer experiences at the time of online purchases. This also empowers the buyer to control the seller's undesirable behavior subjectively and reduce the feeling of risk. Sellers, therefore, promote interaction with buyers to encourage trust-related behavior that helps them in making online purchases.

This study proposes that trust in an SNS can enhance customers' intention to make online purchases. Customers interact on these sites with other customers and online sellers. However, it is the trustworthiness of the site that makes customers believe the review, photographs, recommendations, and promotions posted there. Like the categories of trusting beliefs proposed by McKnight and Chervany [27], trust in SNS also stems from: (a) benevolence, the ability of the site to allow content beneficial for the users, besides seller-promotion details; (b) integrity, which comes from the ethical actions of the SNS and fulfillment of promises; (c) predictability or the uniformity of the content that helps users decide future exchanges, and (d) competence, which is the ability of the site to allow fruitful exchanges, such as evidence of replacement/repair in case of failure from the online vendor side.

9.4.2 ATTITUDE TO SEEK INFORMATION THROUGH THE INTERNET INCREASES THROUGH CUSTOMER'S TRUST

People generally find that information appropriate that matches their knowledge, awareness, and personal presumptions [3] categorizes information-seeking behavior into three types: 1) monitoring, where people observe happenings and other people around them, looking for meanings that can be derived; 2) inquiring, where people ask for information; and 3) consulting, where people access documented information. Information seeking includes customers' activities to gain information about the product they want to purchase from existing resources on SNSes. The information can be in the form of customer ratings, review comments, feedback, and recommendations that can encourage their purchase decisions. Customers look for such information to gain knowledge about the product and its performance. Looking for information also involves appraising the available alternatives and their benefits. Technology helps in cutting down the effort and cost in making informed decisions. Social commerce comprises social advertising, review and rating sites, referrals and recommendations, social shopping, communities and forums, and social media. Businesses use these platforms to enhance customer interaction. Buyers also use these platforms to confirm everyone by various new experiences, thereby increasing the impact of the seller and product.

9.4.3 PURCHASING CHOICES HAVE A POSITIVE IMPACT ON GATHERING INFORMATION

When customers try to acquire information in social commerce, enhanced knowledge helps them in decision-making and increases purchase intentions [39]. Feedback and reviews uploaded by other customers also positively impact sales for online sellers. Using multimedia content, where customers can interact with the products virtually, also has a great influence on the customers' buying behavior. Therefore, information seeking can be seen as a strategy for reducing risks and boosting purchase intentions.

9.4.4 THE PURCHASING POWER OF THE CUSTOMER INCREASES DUE TO SOCIAL SITES

According to Ref. [24], familiarity is defined as "current and/or past use, or knowledge obtained by attending some form of instruction or through readings on the topic." Several studies have attempted to examine the influence of brand or product familiarity on customers' purchase intentions [31]. However, the impact of customers' familiarity with online platforms and their perceptions needs to be further explored. Familiarity with an online platform means the extent to which a customer understands the design and procedures of a site and is confident to share experiences [10]. The higher the familiarity, the lower is the uncertainty. The purchase intention can be described as a technical process that requires sequential steps, such as searching for products, selecting online-vendors, and finding reviews and comments. According to Ref. [9], the complexity in the online shopping process also results in resistance to purchase decisions. In the study [38], it is determined that familiarity with a platform motivates customers to believe in user-generated content.

9.4.5 FAMILIARITY RESULTS IN INCREASING TRUST IN ACCESSING SOCIAL NETWORKING SITES (SNSS)

Familiarity helps customers reduce uncertainty and extend their relationships with others. Familiarity develops based on prior experiences, learning, and interactions [26]. This shows a distinct difference between familiarity and trust. While familiarity understands the current behavior of others, trust is concerned with their future actions [26]. When a user is familiar with a social commerce platform, it represents the user's understanding of the site, its interface, and its procedures. Thus, familiarity develops through the experience of using the platform. On the other hand, an example of trust in the social commerce platform can be providing credit card information; it is a belief that the information is secure and will be used appropriately. Therefore, familiarity is associated with reducing uncertainty by creating a structure [26]; and trust reduces uncertainty by making people have reliable expectations. Familiarity leads to trust.

9.4.6 THE SOCIAL PRESENCE OF THE CUSTOMERS IS INCREASING DUE TO AN INCREASE IN THE PURCHASE

Social presence has its roots in the social presence theory (SPT) given by Short, Williams, and Christie [36]. It highlights that effective communication results from the medium that has a social presence and interpersonal participation. Social presence has been defined as an approach that characterizes the subjective experience of a user with a certain medium of communication. Psychological experts describe social presence as a narrow relationship with intimacy and psychological closeness. Another perspective of social presence is provided [25], which characterized social presence with "perceived warmth, a feeling of human contact, sociability, and sensitivity embodied in a medium" [4]. It is noted that persuasion by strangers is generally more effective. It has been noted that social presence has a close relationship with the richness of information [9]. With a social presence, users can access rich information and meaningful experiences to make sound purchase decisions. Social media platforms and social commerce consider social presence a key requirement. Social presence reduces the social distance between sellers and buyers, making increasing the buyers' trust in sellers.

9.4.7 *SOCIETAL IMPACT ALSO INCREASES BELIEF IN AN SNS PLATFORM*

Social presence is a key feature of social commerce and social media platforms. It is seen as a positive factor in choosing a communication channel, especially in case of an ambiguous and uncertain task, such as online shopping. The social presence of a website delivers richer information and more social cues, and subsequently, such websites are seen to be transparent [25]. High levels of transparency in online environments result in inhibiting untrustworthy behavior [32]. Social commerce lets sellers connect with the customers posting their feedback. A study concludes that the Like button on Facebook, which represents virtual presence, helps customers improve brand assessments and purchase intention [30]. The Tweet button on Twitter and the reaction buttons on Facebook are examples of how social commerce can increase users' sensitivity to relationship quality and social support [8]. They assert that social presence encourages purchase intentions through the aspects of ability, benevolence, integrity, and predictability. Trust in an SNS increases by the presence of others and their observable activities. Facebook allows rich text content, high-quality pictures, and multimedia content; this offers sensitive, sociable, and personal human contact. The recently added feature of adding emojis on Facebook's posts has greatly improved the feeling of warmth in interactions.

9.5 UNDERSTANDING OF SOCIAL MEDIA

Social commerce sites, such as Twitter and Facebook, link customers with online sellers. Not every platform is the same' they differ in their privacy policies, the extent of interaction allowed, various channels sharing information, and amount and category of content which is by friends. This study has chosen Facebook as the target platform because it is amongst the most famous SNSes available across the world.

In addition, it provides tools for posting almost all types of content, such as text, photos, and videos. The study focused on undergraduate students from a private university in Jordan. Students were found to be the predominant respondents for this study. According to Ref. [41], as many as 88% of Facebook users are in the age group of 18–30 years. In this case, almost 82% of users of Facebook have some college experience, and around 79% of students are passed out graduates from college.

Data was collected using an offline survey method (self-administrated/paper-based survey). Questionnaires were developed in English. Only those students were selected who are active on Facebook. A total of 550 questionnaires were posted, out of which 243 were received back, making the response rate of 44.1%. Six questionnaires were submitted with incomplete responses, and therefore, they are deleted. Finally, the result was that 237 people participated in the survey. It is seen that out of a total of 87% of participants, there were 43% females and 57% males, and they all are under the age category of 18–23 years old.

The items to be assessed were adopted from the existing literature (Table 9.1). These things are calculated on a 5-point Likert scale, which is ranging from 1 to 5, wherein 1 denotes strongly agree and 5 denotes strongly disagree.

TABLE 9.1 Measurement Items

Construct	Item
Trust (TRU)	TRU1: Promises made by Facebook are likely to be reliable.
	TRU2: I do not doubt the honesty of Facebook.
	TRU3: Based on my experience with Facebook, I know it is honest.
	TRU4: Based on my experience with Facebook, I know they care about users.
Familiarity (FAM)	FAM1: I am familiar with searching for material on Facebook.
	FAM2: I am familiar with buying material on Facebook.
	FAM3: I am familiar with inquiring about material ratings on Facebook.
Social Commerce Information Seeking (SCI)	SCI1: I use online forums and communities for acquiring information about a product.
	SCI2: I usually use people's ratings and reviews about products on the Internet.
	SCI3: I usually use people's recommendations to buy a product on the Internet.
	SCI4: I trust my friends on online forums and communities.
Social Presence (SOP)	SOP1: There is a sense of human contact on Facebook.
	SOP1: There is a sense of sociability on Facebook.
	SOP1: There is a sense of human warmth on Facebook.
	SOP1: There is a sense of human sensitivity on Facebook.
Purchase Intention (PINT)	PINT1: I am very likely to provide the online vendor with the information it needs to better serve my needs through Facebook.
	PINT2: I am happy to use my credit card to purchase from an online vendor through Facebook.

The points in the survey were prepared and checked by four professors, two from marketing and two from commerce and business, who are related to online shopping, e-business, etc. Their ideas were added before sending out the surveys.

9.6 CONCEPTUALIZATION

Data were analyzed using structural equation modeling (SEM). The method of partial least square SEM (PLS-SEM) was used for this research because it focuses on exploration and prediction; it can manage multi-faceted models, and at the same time, reduces the demands and requirements of data and specification of relationships [34]. This method is considered appropriate for explaining complex relationships, and in the case of investigative and foretelling goals [33]. It is pointed out [34] that PLS-SEM is considered suitable for finding and guessing new models and various theoretical concepts as it works equally fine for investigative as well as current studies. Following, a study [17] which consists of two levels, was used to judge the suggested route of the study model, which involved:

1. The measurement stage; and
2. The structural stage.

9.6.1 *MEASUREMENT MODEL*

The authenticity of the data is calculated at this stage. It is important to calculate: (1) the correctness of individual sets; (2) the different validity of every item is authentic and is calculated by the square upgrading of each set. The square root of the coefficient of each item, which is as per the theoretical concepts, is less than equal to 0.5 [11].

Table 9.2 tells us that all items have squared loading ≥ 0.5. The authenticity of both was calculated by Cronbach's α and combined trust. All the items are having an adequate lot of similarity and are calculated based on Cronbach's α and calculated reliability that shows that all items have squared standardized loading ≥ 0.5. The reliability of each construct was examined by evaluating the coefficients of both Cronbach's α and composite reliability. All constructs acquired Cronbach's α and composite reliability > 0.7 [21]. The coefficient average variance extracted (AVE) was calculated to evaluate the convergent validity [11]. Table 9.2 shows that all the constructs have

AVE value ≥ 0.5, which ensures that every construct tells us a minimum variance of 50% of other related things.

TABLE 9.2 Measurement Model Assessment

Construct	Item	Item Loading	Square Item's Loading (R^2)	α[a]	CR[b]	AVE[c]
Trust (TRU)	TRU1	0.91	0.83	0.92	0.94	0.80
	TRU2	0.92	0.85			
	TRU3	0.90	0.81			
	TRU4	0.87	0.76			
Familiarity (FAM)	FAM1	0.93	0.86	0.90	0.94	0.84
	FAM2	0.90	0.81			
	FAM3	0.91	0.83			
Social Commerce Information Seeking (SCI)	SCI1	0.87	0.76	0.89	0.92	0.74
	SCI2	0.85	0.72			
	SCI3	0.84	0.71			
	SCI4	0.89	0.79			
Social Presence (SOP)	SOP1	0.88	0.77	0.91	0.93	0.78
	SOP2	0.90	0.81			
	SOP3	0.87	0.76			
	SOP4	0.89	0.79			
Purchase Intention (PINT)	PINT1	0.95	0.90	0.87	0.93	0.89
	PINT2	0.93	0.86			

a: Cronbach's a; b: composite reliability; c: average variance extracted.

To find out the areas of this analysis, which suggests that the shared difference between each section, its content must be higher than the shared variance between each section and others [11]. It is seen that discriminant validity was evaluated [7]. The idea suggests that the square root of the AVE for an activity is very high than any relation between the sets.

As seen in Table 9.3, the square root of AVE for a set is higher than any relationship with all the other sets. Thus, discriminant validity is shown in this research.

9.6.2 *STRUCTURAL MODEL*

As seen that after the dataset was decided, the hypothesis was created on its basis. The explanatory power and beta coefficients (β) of the suggested paths are found. Figure 9.3 shows that R^2 for PINT is 0.377, which tells that social impact, trust, finding information, and similarity explains 37.7% of variance data. Furthermore, both FAM and SOP explain 16.8% of the variance in trust (TRU) ($R^2 = 0.168$). Finally, TRU explains 14.9% of the variance in SCI (R2 = 0.149). Regarding the path analysis, TRU ($\beta = 0.262$), FAM ($\beta = 0.168$), SOP ($\beta = 0.298$) and SCI ($\beta = 0.162$) were found to act as facilitators of PINT, as all have positive significant effects on PINT.

TABLE 9.3 Discriminant Validity

Construct	AVE	√AVE	Latent Variable Correlations					
				TRU	FAM	SCI	SOP	PINT
TRU	0.80	0.89	TRU	1				
FAM	0.84	0.92	FAM	0.389	1			
SCI	0.74	0.86	SCI	0.386	0.379	1		
SOP	0.78	0.88	SOP	0.220	0.243	0.192	1	
PINT	0.89	0.94	PINT	0.456	0.404	0.384	0.428	1

Moreover, both FAM ($\beta = 0.356$) and SOP ($\beta = 0.134$) act as important antecedents of TRU, as both have positive significant effects on TRU. TRU was found to be a vital predictor of SCI, as it has a positive significant effect ($\beta = 0.386$) on SCI.

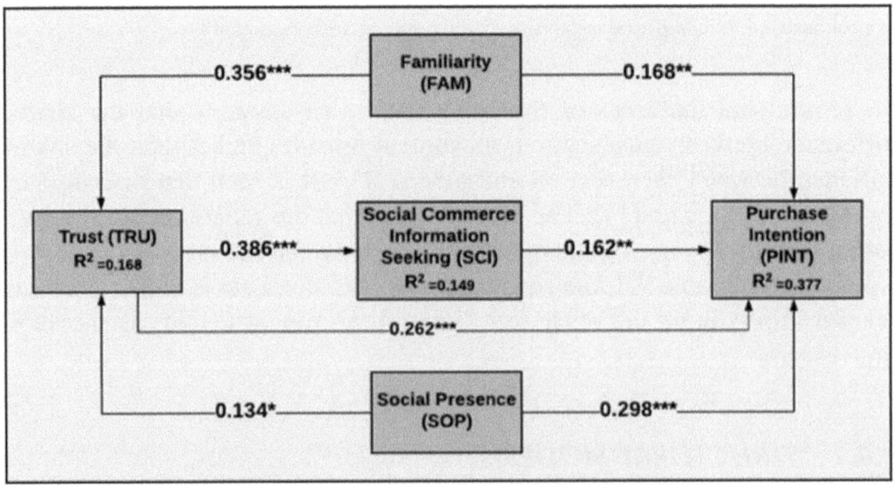

FIGURE 9.3 Result-structural model.

* P-value > 0.05, ** P-value > 0.01, *** P-value > 0.001.

9.7 MEDIATION

A Sobel [3] test was used to show if mediators can significantly convey the effects of independent variables on dependent variables. The results indicated that all indirect effects are important (Table 9.4). It was found that TRU has a significant direct as well as an indirect effect on PINT through SCI. The test proved that SCI extends the impact of TRU to PINT, which highlights the need for the curtail role of TRU in increasing PINT. These findings underline that the increase in TRU can enhance SCI perceptions, and thereby a positive impact on PINT. Moreover, both FAM and SOP have indirect significant effects on PINT through TRU. This clarifies that when FAM and SOP increase; it boosts the perception of TRU, and also leads to a higher PINT.

TABLE 9.4 Mediation Test

Mediator	Path	T-Statistic	P-Value
SCI	TRU→SCI→PINT	4.5	< 0.001
TRU	FAM→TRU→PINT	5.2	< 0.001
TRU	SOP→TRU→PINT	4.18	< 0.001

The research tells us in detail about how to investigate the trust role in the SNS area to influence customers' choices to shop from the online portal through Facebook. It tells an idea to suggest a relation between trust in SNS platform and idea to purchase. This system has three slabs: online social activities, e-commerce information seeking, and a comfortable base. Facebook is an SNS test and research model. SNS has a positive impact on information and purchase, wherein customers with high intentions are very important when the idea of risk and uncertainty is high in e-purchase. Social idea is as important to be reliable. Trust helps in cutting down the impact to see others' activity. Society supports the process of the transfer of information. SNS ideas like various customers' reviews, comments, recommendations, and ratings help people to attract online activity, which has resulted in increased trust with its customers and has reduced risk [5]. Customers are concerned about the word of mouth feedback than information online and try to use all the online sources to access services and products. As discussed by [19], all the information is very positive and is important for e-commerce. It is seen that information also affects brand selection. Customers who read online recommendations are more likely to purchase products online [35]. It is also seen that customers who are sharing their opinion are seeking information by SNS. Customers are communicating through search engines, reviews, comments on an SNS base online to increase awareness by the activities of the website. While many studies have proved that social presence has a positive impact on the customers' purchase intention, some have suggested a negative impact. It also influences trust in an SNS [8]. Collectively, it can be deduced that a strong social presence of an SNS results in a higher number of customer purchase intentions and greater trust in that site. In addition, for online shopping, consumers make a great amount of effort to evaluate how reliable is the information provided by the seller. For this, they take signals from previous buyers. By observing others' purchase actions and reactions reduces their uncertainty. The more transparent the information posted by previous users, the greater is its influence on customers' purchase intention. Social presence also includes the seller's contact with which customers can post their grievances and queries. The details of already addressed queries go a long way than any advertisement in influencing customer decisions. In short, trust in an SNS increases with the social media accessibility of the site (rich text, pictures, and videos), the impact of social media on online sellers (interaction through comments and live chats), and the social impact of other consumers through comments, reviews, ratings, and recommendations.

Familiarity is another factor that majorly affects customers' belief in an SNS platform. It can be described as a facilitator of purchase intentions. The higher the degree of familiarity with an SNS, the greater is the trust and the stronger is the influence on purchase intentions. High familiarity reduces ambiguity in online shopping because it helps customers understand the online environment. Increased familiarity helps in searching for information regarding comments, ratings, forums related to products, services, and sellers. Therefore, customers gain access to more information, which finally impacts purchase intentions. Moreover, familiarity with a site also helps customers find the desired information more easily and faster.

9.8 IMPLICATIONS AND FUTURE WORK

This study makes several managerial and research-oriented recommendations. Managers can use trust as an important tool in social commerce environments. Trust plays a significant role in influencing purchase decisions. The higher the trust in SNS platforms, the greater is the customer engagement. SNS and social commerce developers can boost customer trust by using better features, structure, and design of the platforms. This includes ease of posting and accessing information, quality of information, reputation, transaction safety, and facility to increase buyer-seller interaction (such as live chats). The study also highlights that the ease with which customers can find information on a platform also has a positive influence on customers' purchase intentions. If sellers want to boost customers' purchase intentions, they should enhance access to information and search engines in addition to an improved information exchange process. Forums, feedback, communities, ratings, and reviews also help them interact with peers. Therefore, SNS designers should focus on improving the process of accessing information. Using a search engine facility, optimally can also help in the search process. Studies have also stressed the need to increase familiarity with a site to increase reliability, so changing the look, design, structure, and processes of a site is not a good idea.

This study also stresses the need to make the site user-friendly. Ease of use can help to attract even less familiar customers. In addition, adding various learning means, such as helpful tips, specialized customer support, video demonstrations, and clear guidelines, can go a long way in retaining customers. FAQs, speech-bubbles, and how-to-use tips about the pages are some of the new features being used by various networking sites.

The findings of this study underline the fact that an increase in social presence leads to the higher purchasing power of the customers and their belief in various online e-commerce platforms.

Therefore, sellers, platforms, and social commerce companies should attempt to build and maintain effective IT artifacts to support their social presence. For example, in addition to pages and groups, newer elements, like GIF buttons and animated reactions (as on Facebook), can be incorporated to boost social presence.

As long as the research scope is concerned, most of the previous studies in the area of social commerce have centered around the impact of social presence, familiarity, trust, and exchange of information. Earlier researches have postulated that online surfing through the Internet and similar data are the sources of trust.

However, this study brings forward the concept that this may not be enough in the case of social commerce platforms. This model explores how society has an impact on the purchasing power of consumers.

It also adds the concept of trust as the key element for seeking information, which has a direct impact on purchase intentions. Moreover, it also shows how trust can impact purchase decisions through familiarity.

There are certain limitations in this study, which opens up the scope for future research. Only Facebook was used as an SNS platform. Future research can validate this with other SNS (such as Instagram and Twitter). Moreover, this study has used data that is used to understand the concepts of the model on a cross-sectional basis.

Experimental and longitudinal studies can further be conducted to examine the effect of trust on purchase intentions. This research has used only a quantitative data analysis. Using mixed data collection methods can expose more ideas, facts, and mediators that are responsible for changing the buyer's attitude toward shopping.

9.9 CONCLUDING REMARKS

The research has shown a new method of social commerce and finds put its effects. It also tells us the impact of trust on Facebook in attracting customers to online selling. The research tells us that the difference in other effects and social effects and customers' decisions to e-selling. It is found by the empirical study that online buying increases when the customer is familiar with the website, the extent of its social presence, and the trust it has gained; however, this works only when the customer has social commerce information-seeking behavior. Similarity with the platform also strengthens trust, which adds to the positive impact on online purchase decisions. Trust is a major motivator that drives users to look for information, thereby having an indirect positive influence on purchase intentions. The findings of this study can give actionable insights to website designers, firms, and social commerce managers.

KEYWORDS

- **conducive business**
- **Facebook**
- **online social entertainment**
- **social networking sites**
- **social online sites**
- **web 2 technology**

REFERENCES

1. Ajzen, I., (1991). The theory of planned behavior. *Organizational Behavior and Human Decision Processes, 50*(2), 179–211.
2. Li Yong (2017). Annual report of the United Nations Industrial Development Organization.
3. Busalim, A., & Hussin, (2016). Understanding social commerce. *International Journal of Information Management, 36*(6), 1075–1088.
4. Cialdini, R., (2001). Harnessing the science of persuasion. *Harvard Business Review, 79*(9), 72–81, 29.
5. Do-Hyung, E., Jumin, L., & Ingoo, H., (2007). The effect of online consumer reviews on consumer purchasing intention: The moderating role of involvement. *International Journal of Electronic Commerce, 11*(4), 125–148.
6. Fang, Y., Qureshi, I., Sun, H., McCole, H., Ramsey, E., & Lim, K., (2014). Trust, satisfaction, and online repurchase intention: The moderating role of perceived effectiveness of e-commerce institutional mechanisms. *MIS Quarterly, 38*(2), 407–427.
7. Fornell, C., & Larcker, F., (1981). Evaluating structural equation models with unobservable variables and measurement error. *Journal of Marketing Research, 18*(1), 39–50.
8. Gefen, D., & Straub, D., (2004). Consumer trust in B2C e-commerce and the importance of social presence: Experiments in e-products and e-services. *Omega, 32*(6), 407–424.
9. Gefen, D., & Straub, D., (2003). *Managing User Trust in B2C E-Service, E-Service Journal, 2*(2), 7–24.
10. Gefen, D., Karahanna, E., & Straub, D., (2003). Trust and TAM in online shopping: An integrated model. *MIS Quarterly, 27*(1), 51–90.
11. Hair, G. H., & Ringle, M., (2013). *A Primer on Partial Least Squares Structural Equation Modeling (PLS-SEM)*. Thousand Oaks, CA: Sage Publication.
12. Griffin, R., Dunwoody, S., & Neuwirth, K., (1999). A proposed model of the relationship of risk information seeking and processing to the development of preventive behaviors. *Environmental Research, 80*(2), S230–S245.
13. Hajli, N., & Sims, J., (2015). Social commerce: The transfer of power from sellers to buyers. *Technological Forecasting and Social Change, 94*(1), 350–358.
14. Hajli, N., Sims, J., Zadeh, A., & Richard, M., (2017a). A social commerce investigation of the role of trust in a social networking site on purchase intentions. *Journal of Business Research, 71*(1), 133–141.
15. Hajli, N., (2015). Social commerce constructs and consumer's intention to buy. *International Journal of Information Management, 35*(2), 183–191.
16. Halaweh, M., (2011). Adoption of e-commerce in Jordan: Understanding the security challenge. *The Electronic Journal of Information Systems in Developing Countries, 47*(1), 1–13.
17. Henseler, J., Ringle, C., & Sinkovics, R., (2009). The use of partial least squares path modeling in international marketing. *Advances in International Marketing, 20*(1), 277–319.
18. Indiani, N., Rahyuda, I., Yasa, N., & Sukaatmadja, I., (2015). Perceived risk and trust as major determinants of actual purchase, transcending the influence of intention. *Asian Marketing Journal, 7*(1), 1–13.
19. Ishida, K., Slevitch, L., & Siamionava, K., (2016). The effects of traditional and electronic word-of-mouth on destination image: A case of vacation tourists visiting Branson, Missouri. *Administrative Sciences, 6*(4), 1–17.

20. Kaur, K., & Kumar, S., (2018). S-commerce: An empirical analysis to study the consumers buying behavior. *International Journal of Management Studies, 1*(4), 77–83.
21. Kwong, K., & Wong, K., (2013). Partial least squares structural equation modeling (PLS-SEM) techniques using Smart PLS. *Marketing Bulletin, 24*(1), 1–32.
22. Li, R., Kim, J., & Park, J., (2007). The effects of internet shoppers trust on their purchasing intention in China. *JISTEM- Journal of Information Systems and Technology Management, 4*(3), 269–286.
23. Liang, T., & Turban, E., (2011). Introduction to the special issue social commerce: A research framework for social commerce. *International Journal of Electronic Commerce, 16*(2), 5–14.
24. Liberatore, M., & Titus, G., (1983). The practice of management science in R&D project management. *Management Science, 29*(8), 962–974.
25. Lu, B., Fan, W., & Zhou, M., (2016). Social presence, trust, and social commerce purchase intention: An empirical research. *Computers in Human Behavior, 56*(1), 225–237.
26. Luhmann, N., (1979). *Trust and Power*. Cambridge: John Willey & Sons.
27. McKnight, D., & Chervany, N., (2001). *Trust and Distrust Definitions: One Bite at a Time*. Springer Link. [Online]. Available at: https://link.springer.com/chapter/10.1007/3-540-45547-7_3 (accessed on 1 December 2020).
28. Mesquita, A., (2013). *User Perception and Influencing Factors of Technology in Everyday Life*. Hershey, USA: IGI Global.
29. Mohmed, A., Azizan, N., & Jali, M., (2013). The impact of trust and experience on the intention to purchase in e-commerce. *International Journal of Engineering Research and Development, 7*(10), 28–35.
30. Naylor, R., Lamberton, C., & West, P., (2012). Beyond the like button: The impact of mere virtual presence on brand evaluations and purchase intentions in social media settings. *Journal of Marketing, 76*(6), 105–120.
31. Park, J., & Stoel, L., (2005). Effect of brand familiarity, experience, and information on online apparel purchase. *International Journal of Retail and Distribution Management, 33*(2), 148–160.
32. Piller, F., & Walcher, D., (2006). Toolkits for idea competitions: A novel method to integrate users in new product development. *R&D Management, 36*(3), 307–318.
33. Reinartz, W., Haenlein, M., & Henseler, J., (2009). An empirical comparison of the efficacy of covariance-based and variance-based SEM. *International Journal of Market Research, 26*(4), 332–344.
34. Richter, N., Sinkovics, R., Ringle, C., & Schlaegel, C., (2016). A critical look at the use of SEM in international business research. *International Marketing Review, 33*(3), 376–404.
35. Senecal, M., & Nantel, J., (2004). The influence of online product recommendations on consumer's online choices. *Journal of Retailing, 80*(2), 159–169.
36. Short, J., Williams, E., & Christie, B., (1976). *The Social Psychology of Telecommunications*. London: John Wiley.
37. Technavio, (2017). *Global Social Commerce Market 2017–2021*. Technavio. [Online]. Available at: https://www.technavio.com/report/global-social-commerce-market (accessed on 1 December 2020).
38. Van, D. H. B., & Lim, Y., (2015). On the conditional cueing of credibility heuristics: The case of online influence. *Communication Research, 3*(4), 1–22.

39. Wang, Y., (2017). Antecedents of social network trust in SNS usage: The moderating role of offline familiarity. *Social Networking, 6*(2), 107–134.
40. Yadav, M., De Valck, K., Hennig-Thurau, T., Hoffman, D., & Spann, M., (2013). Social commerce: A contingency framework for assessing marketing potential. *Journal of Interactive Marketing, 27*(4), 311–323.
41. York, A., (2017). *Social Media Demographics to Inform a Better Segmentation Strategy*. Sprout Social.

CHAPTER 10

ARTIFICIAL INTELLIGENCE IN GREEN ACCOUNTING

ANNU PURTHI

Chandigarh University, Punjab, India,
E-mail: anupruthi0911@gmail.com

ABSTRACT

Artificial Intelligence (AI) has revolutionized its application to be the most advanced technology in the world. It helped in extremely critical financial matters, which can assist accountants to better develop a financial strategy and strong client relationships at the same time utilizing their time to help clients in more critical financial matters. This chapter demonstrates how AI is greatly contributing to the accounting sector, specifically in Green Accounting. The proposal of the AI revolution provides ecological balance in society by embracing the challenges of the future and is viewed as a step ahead in the field of development. Conclusions related to green accounting is drawn with the perspective of advanced Intelligence in the accounting field and its applications worldwide.

10.1 INTRODUCTION

It was the initiative of Mr. Jairam Ramesh, the erstwhile Environment Minister of India, who emphasized the importance of implementing Green accounting practices in the country, and the term was coined by Professor Peter Wood in 1980.

Awareness of sustainability practices for the environment has become more vocal in the current century, which is the era of all-round progress globally. Man, in his blind race for development, has caused colossal losses to the environment, and by the time he realized it, the damage was beyond control. The after-effects of this damage have necessitated the evaluation of losses and innovation of new practices to control further damages. So many government and non-government organizations have come out with plans to curb the menace. Everyone needs to understand the relationship between the environment and living beings and spread the message above the masses to contribute to preserving the same.

National Accounting had to be redefined for the reasons that the earlier system (SNA) was found to have a major drawback to the extent that it failed to measure GDP and GNP (gross national product).

Since by now awakening towards preservation of the environment is on the rise, therefore everyone now talks about the prosperity of the same. As a part of the global project, Corporates, and Businesses have joined the mission to define parameters to enhance green and eco-friendly conditions, not only for a comfortable present but also for a safe future for the next generations.

Green accounting, which is also known by many other names like Environmental Accounting/ Resource Accounting/ Integrated Accounting, is just one of the many steps being taken to improve the global conditions about the environment.

Green Accounting Incorporation of costs incurred on account of environmental sustainability into the financial results of operations is called Green Accounting. This would then reflect the sustainability of profits as well as environmental stocks and assets. A revised model for green accounting has become inevitable to reflect the expenditure incurred on saving the environment because the gross domestic product (GDP) hurts this account in the existing pattern.

The conventional accounting model fails to reflect the expenses made for the preservation of the environment or prevention of contamination. It also is unable to represent the cause and effect relationship between an environmental incident and a business or organization. According to Pereira [1], it also does not cover the extent of the exhaustion of environmental resources or take into consideration the degradation of the environment [1].

10.2 THE NEED FOR GREEN ACCOUNTING

1. It is the accounting that factors in economic costs.
2. GDP ion disregards the environment, so policymakers revise to incorporate this.
3. Because of the economic damage to the natural resources of an economy.

4. To make a balance between potential economic goals and environmental goals.
5. Its efforts seek to evaluate the value and status of goods and services provided by the environment.

10.2.1 OBJECTIVES OF GREEN ACCOUNTING

1. **Modification and Detailing of Flows and Stocks of Traditional Accounts Related to Environment:** The expenses incurred for protection and preservation of the environment are arrived at by excluding the stock flows representing the relationship with the environment. Another important objective of this modification is to identify and determine the part of the GDP that might harm the economic process. This is also termed as defensive expenditures.
2. **Establishing a Link between Physical and Monetary Accounts:** The total reserves of natural resources are deemed complete only when these are backed up with a somatic resource account. This account is also inclusive of the changes in the natural stocks caused in due course. We can also say that the System of Environmental-Economic Accounting (SEEA) is the product of monetary stock and natural resource account.
3. **Assessing the Cost and Benefits of Environment:** In a way system of national accounts (SNA) is an expansion of SEEA as it complements the costing structure still further so that we can establish the cost of:
   a. Depletion of natural resource in production and final demand;
   b. The quality changes in eco caused by production-related pollution and other impacts of production, which also obviously covers consumption and natural events and protection aspect of the environment.
4. **Tangible Wealth Maintenance Account:** As we all know, tangible wealth, which is the human-made part of the natural environment, the new concept of SEEA has been designed to cover both aspects. Since it has resulted in changing capital formation into capital accumulation, it helps in the disinterest of environmental assets (Marr, 2018).
5. **Detailed Dimensions of Pointers of Environmentally Adjusted Output and Earnings:** In the process of methodology development, the indicator of environmentally adjusted net domestic product (EDP) has been modified to macro-economic based. This covers the

cost relating to exhaust of natural resources as well as the changes so caused in environment quality.

10.2.2 ADVANTAGES OF GREEN ACCOUNTING

1. Both external and internal users apply green accounting.
2. Green Accounting data makes the basis of investments to be made for ecological expenditure and structure of production.
3. It makes the basis of expenses for ecological purposes while at the same time, the internal debt helps to establish coherence between the environmental expenses and its debt.
4. It also goes a long way in helping in decision making, looking at the plausible implications it may have on enhanced profitability and security of environment basis the analytical data provided about raw materials energy and other information relating to the environment.
5. Accordingly, procurement, usage, and outlets of material and wastage are also well managed.
6. Its contribution to energy and water cost management is another positive factor.
7. It also results in improved affinity between partners and external clients based on the information it reveals about the performance of an economic entity. Such an environmental approach certainly would lead to a better societal image.

10.2.3 FORMS OF GREEN ACCOUNTING/ENVIRONMENTAL ACCOUNTING

1. **Financial Accounting of the Ecosystem:** In this system periodical, financial statements are used to present a true picture of tasks taken up to cover various ecological dimensions.
2. **Management Accounting of Ecosystem:** It covers the aspect of maintaining the records for practices applied for and execution of the same in monetary terms for managing environmental issues. Management Accounting is comprised of strategic planning for environmental management, assessment of benefits derived, and the life cycle aspect as well. Some of the organizations also include the auditing part of this accounting in their reports.

3. **Cost Accounting of Ecosystem:** It is the process of an in-depth inclusion of expenses incurred on account of various environmental aspects such as identification, evaluation, and allocation of conventional costs. Not ending here, they also account for social costs to processes, products, activities, or budgets. We must understand that the processes of dealing with pollution may also result in emitting some amount of pollution. Therefore, going by the accounting polluter pays principle (PPP), the organization responsible for the pollution produces will have to pay for dealing with that pollution. In case he fails, this may lead to external environment costs, which means a third party will have to take this responsibility on account of the failing party.

   The two major dimensions of the term environmental costs are:
   a. Discerning the directly impacting part of it over "private costs"; and
   b. "Social costs," which include individuals, society, and environment together, for which a company can't be held accountable.
4. **Ecological Accounting:** This term deals only with the physical information used for the preparation of accounts. It also covers Natural Resource accounting at the local administration level. This kind of accounting system is designed to prepare and plan the management of eco-related assets at the administrative level. These plans help in establishing the life cycle of the system under study under given conditions and explicit physical quality.
5. **Accounting System for Natural Resource:** This particular system refers to the inclusion of environmental aspects into the SNA with a specific focus on the quality deterioration of natural assets. This, they feel, would lead to environmentally adjusted economic indicators like environmental gross national income.

10.2.4 INITIATIVES WERE TAKEN FOR GREEN ACCOUNTING

Steps were taken in India in this regard: Under the Companies Act of 2013, the government of India brought all companies with:

a. The net worth of Rs. 500 crore;
b. Turnover of Rs. 1000 crore;
c. Net gains of Rs. 5 crores.

Under the preview of making them socially responsible for the virtuous job of environment protection by enforcing Corporate Social responsibility (CSR) clause to pay at least 2% of their average net profits for the last 3 years towards CSR activities, which figures also have to be reflected in their annual Board/Directors' Reports.

Disclosure of expenses incurred and work done on energy conservation, technology development, foreign exchange earnings were also declared mandatory by the government for all companies falling under the CSR category.

However, the way the needs for environmental conservation and protection of businesses are rising day by day, the above-mentioned steps may not suffice the purpose.

10.2.5 STEPS BEING TAKEN IN OTHER COUNTRIES

Green Accounting Practices are commonly being adopted across the globe because the developed nations have realized the need for conservation of energies to ensure a better future for the next generation. Australia is the first country to initiate the Australian Institute of Environmental accounting, which assumes the responsibility to educate people through a structured course on sustainability and Carbon and Energy and Environmental Management in a row with Germany, the U.S., the U.K., Spain, and France.

In line with the Companies Act of the U.K., which ensures enrichment of companies' efforts, The Directors of such companies falling under CSR have to, compulsorily include information in their reports regarding the kind of environment they work in and the risks involved. The French government too is very strict in the implementation of this act, while it is the responsibility of the French Ministry of Sustainable Development to publicize the environmental-economic accounts every year.

10.3 APPROACHES ADOPTED IN GREEN ACCOUNTING

It was inferred by Rewadikar [2] that Environmental Accounting is based on two approaches, namely Physical and Monetary. To arrive at desired information relating to sustainable development, monetary data alone is not enough because it can explain the value relations of the elements between the economy and natural assets. So, it, therefore, must be complemented by physical data:

1. **Physical Approach:** It is to include physical assets into the SEEA (system of environmental and economic accounting) asset balance sheet. For example, opening and closing stocks of material, agriculture, desert land, natural resources, and the changes caused in line with their usage and change of state. Simultaneously, environmental quality changes in natural resources also form a part of this work. In other words, the environmental operations are expressed in physical terms like balance in hand, additions, and deductions purely in terms of quantity without assigning any monetary value.
2. **Monetary Approach:** It is concerned; it talks about the funds spent on the protection of the environment as given in national accounts. The advent of this approach resulted from the insufficiency of the physical approach system, which talks only about assets and not asset values. Although the monetary approach is not so easy to adopt, it is required to arrive at profit and loss associated with environmental operations. It has also made it possible to induce environmentally adjusted economic indicators.

10.4 FUNCTIONS OF THE GREEN ACCOUNTING

1. **Internal Functions:** As one step of a company's environmental data system, internal functions make it attainable to manage environmental conservation expense and analyze the expense of environmental conservation activities versus the profit obtained, and promote effective and efficient environmental conservation activities through suitable decision making.

   It is fascinating for environmental accounting to function as a business management tool to be used by managers and connected business units.
2. **External Functions:** By revealing the countable measured results of its environmental conservation activities, external functions enable a company to impact the decision-making of stockholders such as-consumers, business, partners, investors, residents, and government.

10.5 AGGREGATION AND VALUATION OF GREEN ACCOUNTING

National environmental accounting is an adjudged basis for environmentally adjusted (rates 'greened') indicators, which include a sum-up of inputs, outputs, and environmental impacts. SEEA formulates this basis prices and costs (monetary aspect) and the mass of material flows (physical aspect in terms of weights). Environmentalists, however, by Yarahmadi and Bohloli [3] discourage the basis of calculating the value of these resources in terms of market rates because they rightly believe that these are an invaluable gift of nature and hence cannot be treated as any other commodity. That is the reason the terminology 'aggregating material flows' has been used for the physical aspect in material flow accounts, which signifies the consumption rates as an alarm to arrest wastage. At the same time, it needs to be clarified that these diverse environmental impacts relating to depletion of resources also signify the adverse impacts of the same on the environment and the planet life.

However, paradoxically most researchers have based their studies on green accounting based on market rates or obtainable/ natural resource rents for non-produced natural assets. Obtainable/ Natural resource rents refer to the money earned by selling resource outputs in markets. This makes the base for establishing the net prevailing price of a resource output linked to the estimated value change of an asset due to the depletion of the resource. For environmental degradation or damage evaluation, expenses incurred for palliating or eradicating the environmental impacts are used to establish results.

10.6 POLICY USES OF GREEN ACCOUNTING AND DIFFICULTIES ATTACHED TO IT

Green accounting indicators are just not used as performance data or assessing the sustainability of an economy's performance over a given accounting period. Rather, these indicators are helpful information about policies and evaluation of results attained.

The key element in establishing the growth of an economy is the enhancement of productive wealth. Natural resource rents in resource-rich developing countries, which form a part of natural resource accounts, maybe amalgamated in development funds, which could be further utilized for long-term development projects, thus saving the earnings.

The ecological costing incurred by economic agents of personal or commercial enterprises is one of the unique strength areas of green accounting for which the agents are principally held responsible for accounting for the environmental impacts of investment by polluter/users. The economists conceive that globalization of instruments of the environmental cost will be a better idea to efficiently attain sustainable production and consumption patterns compared to the prevailing environmental regulations. In a situation wherein required information about green accounting is unavailable, most of the market instruments settings are determined by political extremity, which otherwise is desirable to be based on rational cost estimates.

A big challenge for the current generation of the planet is to draw a balance between usage and sustainability of resources in such a way that while we can meet our requirements of the current time, we also must take responsibility to ensure that the future generations are not deprived of their share.

Though the planet abounds in 'matter' (what we call resource), all resources may not be of any value for certain species. Say, for example, carbon dioxide is of no use for breathing, and what shall we use volcanic lava for. Hence resources are of subjective valuation; hence we also have to be subjective while planning equilibrium between usage and sustainability.

The equation, however, is not as simple as it seems because subjective valuation leads to a complicated scenario where one has to correlate the resource with the potential user. The complexity of the problem is unmanageable even with the help of rocket science, the terminology we use for difficult things. Since man is not capable of designing a perfect system like the one nature can, for example, the way we inhale oxygen through the air, which is a mix of gasses some of which could be injurious to health or else fatal for man, we need to chalk out better and efficient projects to make it happen to the optimum.

Some opine that planning things at a central level by including all determinants that fulfill the human desire and/or provide them some sort of happiness is a better way to deal with this enigma. However, this too is merely a statement that is neither supported by a logical solution nor any univocal justification. Those who advocate this philosophy favor such central planning only because they hope that this approach could optimize the human happiness element. However, the conflict remains because the other part proponents disagree with this philosophy. So it leaves the question open whether the decision is left to class controlling the market, or should the planners work on it?

10.7 GREEN NATIONAL ACCOUNTING

With so many scuffles around this point, the UNEP and the World Bank commenced a row of seminars during the 1980s to find an alternative to this most debated determinant of economic growth. As a result, the UN was able to come out with the most developed and vigorous effort was made to find an alternative to national accounts resulting in 'System of Environmental and Economic Accounts' (SEEA), and thereafter, 'Integrated System of Environmental and Economic Accounts.' With all its strengths, it seems the most acceptable tool till now.

The changes in natural resources should not be integrated with the framework of the traditional SNA but captured separately by creating a separate "satellite" account.

Or integrate the measures with the traditional SNA—although this would only be restricted to easily valued resources (oil, coal, and timber) and would not take into account other ecological factors, such as pollution.

IEEA was initiated somewhere in 1990 in Latin American countries such as Colombia and Mexico on an experimental basis, which would run parallel to the existing SNA. If it is found satisfactory, the IEEA might replace the traditional SNA in times to come. The IEEA green accounting is expected to bring to light some new indicators to replace the traditional GNP or GDP. For example, they talk about the eco-domestic product (EDP), which means mining activities will have a different impact than they have in SNA. For example, in case of an increase in the export of minerals, it will show a negative impact on mineral resources contrary to increased GDP under SNA. The negative figure will suffer the profits arising from the production and export of the resources.

Problems are arising from adopting EDP:

1. Since EDP amounts to double-counting, it tends to double count the sunk cost of capital investment. This, in turn, would lead to twice as much tax on the number of minerals extracted and hence the cost mineral would rise high impacting all downstream activities which are interdependent on them.
2. While the nature of resources keeps changing with time, in this scenario, EDP, which is dependent on a static, would be constrained.

It has been truly observed and reported by the UN that many countries give a fake valuation of income and wealth because the value of resources and environment depletion is generally ignored by the traditional SNA. As the UN says, "Incorporating the real value of natural resources as well as their depletion and degradation allows for better allocation of priorities, thereby helping to address the causes of current major environmental problems including the over-exploitation of natural resources such as forests." It may not be possible for the government officials to do it as they are unaware of the evaluation process of such resources by the individuals. In other words, these prices are based on the market place where people enjoy the right to sell or purchase these things, and the values are decided based on their choices of the product being bought or sold. The SNA incorporates the subsequent choices.

10.8 PROBLEMS OF GREEN ACCOUNTING

Numerous problems are likely to be excruciated with the Green Accounting method, which are:

1. Since the natural resource accounts which affect a particular region do not form a part of the main Green Accounting, it can't be called a compendious natural resource accounting.
2. The focus remains limited to the usage of these resources for economic activities; the transformations and flow of activities happening within the natural resources do not figure anywhere.
3. Another constraint has been the unavailability of the variety of data needed in a required format; hence the accounting remains incomplete.
4. Another constraint is the allocation of environmental pollution heads to the related economic activities because as the causes of pollution are ambiguous and anonymous, the cost of pollution prevention remains unidentified. The matter is further complicated as the causes of pollution are of different types; there are no clear indicators on how to assign the cost of damages. Therefore, connecting environmental data with existing national accounts data for preparing Green Accounts represents a false image.
5. Also, some of the environmental pollution consequences do not come to the surface instantly; hence formulating policies basis, visible consequences would lead to wrong decisions.
6. While SNA is the simplest where market prices are directly applied, different valuation bases such as prevention and restoration costs and contingent evaluations based on surveys are used for different aspects of environmental problems in the System of Environmental-Economic Accounting (SEEA). In other words, this is not a justifiable system because there are mainly theoretical and arbitrary constructions in SEEA.
7. The pricing of all environmental variables in monetary terms in the Green Accounting has consequences:
    a. Green accounting has a restricted accounting system as it deals with only such variables that are easily monetized, and as such, it

limits the range of the accounting system. All this is consequent to the pricing of all ecological factors in money terms.

b. The experiences with the Green Accounting system have compelled man to device some sort of artificial intelligence (AI) program because coining into the money of environmental malleable and its immersion of only a few aggregates results in a severe reduction of the System of Environmental-Economic Accounting (SEEA) system.

An acceptable Green Accounting approach must include the viewpoint of those who believe that 'development' means way more than 'economic development.' It ought to be, inside the boundaries of practicability, commit to showcasing the complete set of goods and services that we acquired from worldwide. It should be structured to include factors valued by those advocating a multiplicity of perceptions of sustainability and factors advocating a variety of perceptions of outsiders. It should mirror the considerations of those who view themselves as being or turning into worse off in conjunction with those who see themselves as favored.

After knowing all the pros and cons of Green Accounting, there is a growing trend of AI.

In the accounting profession, where people deal with rote tasks to achieve more efficiency, AI a replacement of human capabilities for the better. It not only helps to speed up payment processing, invoicing, and stock pupations due to automation but also efficiently keeps track of purchasing and order status. r. It is being used by many accounting firms where it analyzes a large volume of data at high speed, which would not be easy for humans.

10.9 AI AND GREEN ACCOUNTING

AI is a human-made brainpower process by machines, particularly computer systems, which incorporates reasoning, learning, and self-correction.

AI application includes expert systems, machine vision, and speed recognition. AI is taking the accounting services industry by storm. For example, Robo-advisor Wealth front tracks account activity using AI capabilities to research and perceive how account holders pay, invest, and built monetary choices to customize the advice they give to their customers.

The accounting profession has existed since prehistoric times. During its long journey, it has seen many transformations as a result of the changing world and the resources available.

Accounting software exhibits superior performance as compared to the ancient methodology of pen-paper-based accounting. This evolution of technology led to the digitalization of the entire accounting process.

This software uses AI capabilities to automate tasks like data entry, reconciliation, account payable, and many more.

AI helps in extremely critical financial matters, which can assist accountants to better develop a financial strategy and strong client relationships at the same time utilizing their time to help clients in more critical financial matters. If AI needs no human intervention to work, it is not a point of worry for accountants because they just need to adapt themselves to this new method of working and accept the change with an open mind. After all, it would lead to a better future.

According to Marr [4], "Automation, miniboss, machine learning (ML), and adjustive intelligence have become a part of the finance team at lightning speed."

This statement vouch for the facts we have been talking about in the foregoing pages that with the growth of ML and AI the human beings are bound to be more productive and proficient because these applications consistently continue to uprise and upgrade accounting responsibilities as well as the growth of human professionals. With the inclusion of such machines, a more innovative approach will be adopted, which in turn would enable humans more of client building and deliver more valuable results. These intelligent machines are not just tools but rather colleagues of human beings at the workplace, assisting them in performing bank reconciliations. At the same time, the emotional quotient remains intact because not all is left to machines; men still would participate with them.

10.10 IMPACT OF ARTIFICIAL INTELLIGENCE (AI) ON GREEN ACCOUNTING

Kadian [5] discussed a few implications of AI on green accounting, as stated below:

1. Producing result-oriented data:
    a. Providing valuable and analyzed information to decision-makers of accounting firms.
    b. Automating the approval of workflows that further helps to limit the risk in the industry.

2. Calculating risk assessment and bonus accruals:
   a. Machine functions on system information and predicts bonus accruals;
   b. Analyzes previous projects and reduce total risk on the system.
3. Transactional accounting task:
   a. Simplify the auditing process of expense claims;
   b. Intelligently clearing the invoice payments and generate an invoice.

10.11 BENEFITS OF ARTIFICIAL INTELLIGENCE (AI) IN ACCOUNTING

1. AI helps to automate tasks like payment processing, invoicing, and updating stocks.
2. AI can keep track of the status of orders as well as purchasing.
3. It helps accountants to segregate data into the corresponding categories.
4. It can do critical task audits, tax filing, payroll, and baking.

10.12 ACCOUNTING TASKS MACHINES CAN DO

This is juncture leading to a stormy transformation of accounting and finance with the help of digital technologies taking shape at a much faster speed and all this blended with more mature tech capabilities will help man to store and success financial information much more easily than by manual means. In the event of the 4th Industrial Revolution [4] tiptoeing soon the taking over of human jobs by machines in no way is going to diminish the job prospectus of accountants because in the event of machines taking over the tedious jobs of repetitive, time-consuming, and redundant tasks the finance professionals will get more time to spend on analyzing the financial scenario and guiding their clients to better investment opportunities. Let's have a look at a few opportunities available due to AI. Thanks to the digital transformation of accounting and finance.

The automated mechanisms are more efficient in operations, and hence the cost of operations will be reduced due to less time is taken, and errors avoided. This will give humans an increased chunk of free time. This will further help them to gain enhanced expertise that will add value to their future business process transformations, and they will end up being financial leaders.

10.13 ACCOUNTS PAYABLE/RECEIVABLE PROCESSING

We owe our thanks to digital workflows, which have gone a long way to do our job not only easier but also efficient and accurate. For example, AI-powered invoice management systems, which are already in place, have made invoice generation a completely streamlined process as the system can learn accounting codes much better than humans.

10.14 SUPPLIER ONBOARDING

These machines have the capability of investigating thoroughly the new suppliers and decide if they should be accepted basis their credit scores or tax information. If found acceptable, they also include them in the system without man intervention and further set up inquiry portals so that the required information is obtained about them.

10.15 PROCUREMENT

The purchasing process of organizations is very cumbersome for organizations as it involves a lot of paperwork. Since different systems are used for the same, the files are not congenial and agreeable with one another. The machine, on the other hand, helps to make this tedious job paperless because machines can integrate the unstructured data by processing it through APIs. Robots have been identified to be ideal to track price changes across numerous suppliers with extreme efficiency and accuracy.

10.16 AUDITS

Even the audit process has been revolutionized with the advent of digitization because it additionally provides desired security measures since the digital trail is recorded each time a file is accessed with the identity of the person. This also makes the audit process simpler and smoother because the auditors don't have to search for physical files required by them. This job is done by clouting the digital files.

More digital audit improves efficiency and effectiveness. It results in the accuracy of audits and makes a cent percent audit of a company's financial transactions possible instead of just a sample.

10.17 PERIODICAL CLOSE PROCESS

Machines can consolidate and reconcile data that is posted from numerous sources making all the periodical (monthly, quarterly, half-yearly, or annual) closing process not only easier but also speedier, accurate, and first time right.

10.18 EXPENSE MANAGEMENT

Machines can also read bills and receipts for reimbursement of expenses as well as audit the payments made to this purpose, which otherwise is quite a time-consuming job and usurps valuable time of the accounting team. Machines also raise the alarm in case of contravention of company policies for any payment made.

10.19 AI CHATBOTS

To find an efficient solution to some common questions and queries from the customers, Chatbots are generally used, which even discern account balances, account statuses, and even the due bill payments.

Keeping in mind the fact that machines are efficient inaccurate analysis of data and can also read patterns in the data so that these can decide a judicious way of data usage, it is essential to pair people and machines, which is the entry to the digital transformation of accounting and financing. This way, both men and machines can be allowed to contribute to their proficient areas. Since machines are capable of mind-numbering and tedious tasks, they will help financial professionals to take up tasks they are more proficient in.

10.20 CONCLUSION

Most of the audit tasks, which are very complex in nature and style, have been carried out by accounting researchers without the involvement of AI experts. However, it is believed that there exists a big scope of improvement in research on AI if they work together with accounting researchers. Another area wherein the firms, organizations, and accounting departments and the computer science and AI domains need to focus on is bridging the gap between the two, which should improve auditing and assurance. Expert system technology forms an essential part of most of the AI research in accounting, finance, and especially auditing. Therefore, more complex AI applications can be created to solve auditing problems more accurately and with compatibility. Many of the complex audit and assurance problems have been dealt with techniques like expert systems, neural networks (NN), fuzzy logic, and other ways of AI.

Tasks like assessments, review procedures, risk calculations, and going-concern decisions, which are essentially audit works, are complex but important. Performing these tasks poorly has adverse consequences. The potential for improvement through the up-gradation, development, and use of complex AI applications, such as genetic programming, NN, expert systems, fuzzy logic, and hybrid systems, should be investigated to the full possible extent.

KEYWORDS

- **artificial Intelligence**
- **eco-domestic product**
- **environmental friendly accounting**
- **green accounting**
- **polluter pays principle**
- **sustainable development**

REFERENCES

1. Pereira, L. J., (2017). *What is Green Accounting and its Importance.* https://www.annualreviews.org/doi/10.1146/annurev-environ-042911-093511 (accessed on 1 December 2020).
2. Rewadikar, B., (2014). *Advantages of Implementing Green Accounting.* http://www.indianresearchjournal.com/wp-content/uploads/2014/08/ADVANTAGES-OF-IMPLEMENTING-GREEN-ACCOUNTNG.pdf (accessed on 1 December 2020).
3. Yarahmadi, H., & Bohloli, A., (2015). *The Effect of Green Accounting in the Society.* http://article.sciencepublishinggroup.com/html/10.11648.j.jfa.20150305.15.html (accessed on 1 December 2020).
4. Marr, B., (2018). *The Digital Transformation of Accounting and Finance-Artificial Intelligence, RobotsandChatbots.* https://www.forbes.com/sites/bernardmarr/2018/06/01/the-digital-transformation-of-accounting-and-finance-artificial-intelligence-robots-and-chatbots/#6135586d4ad8 (accessed on 1 December 2020).

5. Kadian, N., (2017). *Implementing AI and Machine Learning into Accounting.* https://www.slideshare.net/mobile/nishantisation/implementing-ai-machine-learning-into-accounting-77690021 (accessed on 1 December 2020).
6. Mondal, P. (2014). *Green Accounting: Need, Objectives, Problems and Other Details.* http://www.yourarticlelibrary.com/economics/environmental-economics/green-accounting-need-objectives-problems-and-other-details/39675 (accessed on 1 December 2020).

CHAPTER 11

IMPACT OF ARTIFICIAL INTELLIGENCE ON THE INSURANCE INDUSTRY

RAVI SANWAL

S. V. Government PG College, Lohaghat–262524, Uttarakhand, India, E-mail: ravisanwal6@gmail.com

ABSTRACT

Insurance is an old and profoundly directed industry. Maybe along these lines, Insurance agencies have been slower to grasp mechanical change contrasted with different enterprises. In today's digital world, insurance is still far behind to create new services incorporated using AI. Even today, there are lengthy paperwork procedures, and customers do face problems of slower processes when getting their claim reimbursed or framing in a new Insurance policy. The expanding capacities of man-made brainpower (AI) are changing how associations work and collaborate with clients both inside and externally. The insurance division is right now utilizing AI in a few different ways, yet its capability to upset insurance is not clear.

With chatbots for faster customer service and with machine learning (ML) technologies that function using an algorithm with structured data, AI enhances the performance of any workplace with its automation process. With increased awareness about the reformation of processes through AI in the insurance industry, the initial reluctance towards its implementation is now declining at an increasing pace, as it has spread its roots wider for newer opportunities brought about by ML and deep learning.

Man-made intelligence's underlying effect will identify with improving efficiencies and mechanizing existing clients confronting guaranteeing and guarantees forms. After some time, its effect will be increasingly extensive; it can recognize, evaluate, and endorse developing dangers and distinguish new income sources, affecting each part of the P&C Insurance industry.

11.1 INTRODUCTION

Insurance is a kind of risk management wherein the insurer bears the burden of potential loss required to be legally paid, incurred by the insured that undertakes the Insurance risk for compensation called a premium. Insurance protects individuals, businesses, or any other entity against unwanted, unavoidable, inevitable financial losses (up to a certain limit) based on some uncertain future event. The bottomry contract was one of the oldest insurance. Insurance is an agreement or an integrated contract between insured and insurer based on certain terms and conditions, termed as an Insurance policy on which both the parties agree upon. Insurance Regulatory and Development Authority of India (IRDAI) is the autonomous regulatory body empowered with regulating the workings and functions of Insurance (Life and Non-Life insurance) and the reinsurance industry in India [1].

Simulated intelligence makes machines wise. It is an umbrella term given to a heap of innovative headways that have helped people to program machines to play out specific exercises. Man-made intelligence has become incredibly well known as a result of its capacity to robotize undertakings performed by people and finish them in a limited ability to focus time without exhaustion or mistakes. To see how back up plans can best use the capacities of AI, you should initially take a gander at the desires of a cutting-edge purchaser. Everything today should be quick; however, it should likewise be truly close to home. Customized encounters are the new norm; however, conveying them at current market costs would be inconceivable without AI [2]. With AI in the Insurance business, back up plans can make extraordinary encounters for clients hoping to purchase or re-establish property and loss Insurance. Everything from cases to guaranteeing information investigation can be affected, streamlined, and enhanced to improve client assistance and interior activities.

Insurance companies are leveraging artificial intelligence (AI) to solidify their foundation in today's customer-oriented digital world. AI has enhanced and evolved over the years to match in with the dynamically changing customer preferences and scenarios. There is a global push by Insurance industries to add to technological advancement for faster, cheaper, and digital business initiatives. AI has become extensively favored because of its ability to handle human tasks in an automated manner and complete them in a short period without errors. In an age of online activities, digitized, and convenience is the probable reason to influence the whole of the Insurance industry. AI can lead to effectiveness and timesaving by automating the labor-intensive processes-may lead to a happy customer experience. Insurance companies can analyze and anticipate consumer behavior, better understand customer preferences, and offer better product offerings. Today, with the dynamically changing scenario, an introduction to autonomous vehicles could bring the most significant change in the Insurance industry. KPMG has been at the frontlines of this transformation, with *"Self-Driving Cars: The Next Revolution."* In a 2015 report, KPMG said the Insurance market would shrink by as much as 60% by 2040.

11.1.1 AI AND INSURANCE

Like each other field, the protection business is utilizing AI to streamline its tasks, draw in with the crowd in a powerful way, and encourage deals. Simulated intelligence has tested inheritance product offerings as well as upheld InsurTech organizations to earn funding. Principally, AI has set aside a great deal of cash for the safety net providers and has made life less complex for the policyholders [3]. Computer-based intelligence can help mechanize work serious procedures, prompting lower costs and spared time. AI can likewise be utilized to comprehend clients better—organizations can utilize AI to dissect the information they have on clients to foresee client conduct, get inclinations and streamline cost and item contributions.

Computer-based intelligence is included many related advances, some of which are:

- **AI:** Includes preparing PCs to distinguish designs in the information or potentially foresee results. Other AI advances are utilizations of AI. AI is frequently used to create quantitative exchanging procedures.
- **Profound Learning:** It is a revolutionary technology of deep learning and AI wherein the model breaks the information to make the inferences and handle the issues without any expressed guidelines. This function emulates the human brain in analyzing data and understanding patterns for decision-making.
- **Neural Systems:** It is either a system software or hardware that functions or imitates the workings of neurons in the human brain, and perceives patterns in decision-making. Neural networks (NN) consist of both input and output layers to recognize and view various

patterns. These are the tools to recognize patterns that are complex for programmers to code. An example of profound learning and the neural system is Google's QuickDraw, an online fun game to recognize doodling and to guess what the drawing represents. This AI reads from each data set of drawings, recognizes its pattern, and increases its ability to guess correctly in the future.
- Normal Language Preparation: It enables PCs to comprehend, decipher, and react in composed content or discourse. This tech is regularly utilized by chatbots.

11.2 SUBSETS OF AI

- Man-made intelligence calculations are utilized to group and study information and distinguish connections [4].
- At the point when applied to informational collections, AI can be utilized for design acknowledgment, improvement, and forecast.
- AI can characterize and investigate information in various arrangements: content, discourse, picture, video, and so on. It can likewise work with organized (e.g., marked information) and unstructured information.
- AI calculations learn by being sustained by huge informational collections of named information. When they can distinguish the right ends from the realized informational index, they can be applied to true issues.

11.3 THE NEED FOR AI IN INSURANCE

The insurance company majorly faces a challenge in three main areas; firstly in reaching out to potential customers; secondly providing the requisite solutions that meet the customer's preferences; and thirdly, at present insurance company, faces the challenge with quick claims and identifying and rejecting the fraudulent claims by customers. Insurance companies particularly are far behind in investing in AI-enabled technology that can not only reduce the tedious manual administrative work but can also enhance the service quality and experience for customers by making the right decisions [5]. The technologically advanced system of AI keeps the agents and employees synchronized, which makes the work a little more transparent and quicker decision-making.

11.3.1 HOW AI WILL IMPACT THE INSURANCE INDUSTRY?

1. **Underwriting:** The endorsing procedure relies upon information on the board and investigation, a procedure generally supervised by individuals. With the approach of AI, the examination of endorsing criteria can be handled quicker, creating precise evaluating results and hazard appraisals. It is not necessarily the case that the human component is expelled from the endorsing procedure the exact inverse. Reports made with AI can educate financiers and their basic leadership forms, making their employments simpler and their examinations progressively exact. Man-made intelligence in protection impacts the accompanying regions of the endorsing procedure:
   - **Pricing:** AI can assist guarantors with taking a gander at present and past hazard levels and evaluating rules that connect. This procedure helps P&C back up plans even more precise value approaches inside a fitting window, without "leaving cash on the table" [6].
   - **Risk Management:** AI can produce hazard the board steps dependent on noteworthy information during guaranteeing and would then be able to screen arrangements and insureds during the hour of inclusion. This can improve claims expenses and associations with clients.
2. **Claims:** Cases have generally been a drawn-out procedure. In addition, when a client is compelled to record a case because of damage or property harm, the exact opposite thing they need to do is manage different purposes of contact. This is the place AI can help. It can empower guarantors to best serve their clients in their critical crossroads. Cases handling programming fueled by AI can process claims, appoint a specialist, evaluate inclusion levels, make follow-up assignments, and remain in contact with the client all through the case procedure. This diminishes process durations and gets clients the assist they with requiring rapidly, improving their associations with their safety net providers and their backup plans' notorieties in the market [7]. Computer-based intelligence is likewise being utilized to battle an immense cash pit in the protection of false cases. Smart frameworks can spot abnormalities or examples in information and ready back up plans for conceivable extortion.
3. **New Business:** Man-made consciousness in protection is imperative for new business, as it is helping back up plans give customized

encounters to clients with no underlying human cooperation. Utilizing customer-created information and AI calculations, safety net providers can offer best-fit approaches to forthcoming clients dependent on their one of a kind pre-requisite. Computer-based intelligence can even prescribe inclusion levels dependent on past client collaborations or purchasing practices of clients who fit comparable profiles. The weight of distinguishing the ideal item is no longer exclusively on a specialist and is rather intensely upheld by huge measures of information that must be handled utilizing man-made brainpower.

4. **Retention:** AI can likewise assist guarantors with forestalling weakening or passes with their present clients. These endeavors all beginning and end with information and the experiences they create on clients, their practices, and their hazard for allowing their approaches to policies. AI can process information that backs up plans that have gathered just as information outside of their frameworks. Web-based life data, socioeconomics, buy history, ecological components, and more would all be able to paint an image of a client. Calculations can utilize this information to distinguish clients needing consideration and empower you to concentrate your re-commitment battles on that client set. At that point, utilizing AI, you can dispatch a re-commitment battle that contacts clients through their channel(s) of decision and when they'd be generally responsive to contact. Simulated intelligence commitment for clients can be upgraded and used to choose legitimate channels, suitable items, focusing on, and timing of informing to be best.
5. **Marketing:** To remain focused in the protection business, safety net providers need to grow new items now and again. In any case, that exertion is in support of nothing if the items don't sell, putting forth promoting attempts essential to progress. Like maintenance endeavors, AI can likewise educate showcasing endeavors to target clients most adequately [8]. The computer-based intelligence information can be utilized to create profiles for client gatherings, which advertisers would then be able to use to advise and coordinate their systems. Simulated intelligence can likewise screen leads as they work their way through the business pipe from attention to enthusiasm to choose to activity and afterward direct focused on showcasing endeavors to these potential clients at each phase simultaneously.
6. **Improving Operations:** AI in protection can improve all aspects of approach organization, endorsing, charging, and client relations;

however, it ought to likewise be utilized to improve the inward activities of guarantors themselves.

AI makes safety net providers progressively proficient. Clients can be reached and adjusted at a higher volume and higher speed, setting aside back up plans cash and making more joyful clients. It can likewise accelerate claims handling via computerizing basic leadership forms and getting clients the cash they need.

11.3.2 ARTIFICIAL INTELLIGENCE (AI) AND CUSTOMER EXPERIENCE

The entirety of the previously mentioned zones of the protection business that can be improved by AI—guaranteeing, claims, new business, maintenance, advertising, and operational efficiencies-are planned for improving the experience for clients. Toward the day's end, consumer loyalty is the essential worth include of AI in protection. Clients hope to be a need previously, during, and after the business procedure, a degree of consideration that would debilitate almost every asset for back up plans. With AI in the protection of client assistance, organizations can take care of each need of their clients at the speed and recurrence that is required.

To be fruitful, AI relies upon one asset-seemingly the most significant asset for P&C back up plans-information. What's more, for safety net providers to effectively use the intensity of AI, they need an information investigation arrangement that can gather, break down, oversee, and report on the information they're gathering each second [9]. When an appropriate arrangement is set up, AI programming can be given something to do, consequently utilizing information to assist guarantors with settling on the best choices for their organizations.

McKinsey gauges a potential yearly estimation of up to $1.1 trillion if AI tech is completely applied to the Insurance business [10]. Of this, the business zones that can profit the most are:

- **Sales and Marketing:** AI can be utilized to value protection arrangements even more aggressively and pertinently and prescribe helpful items to clients. Back up plans can value items dependent on singular needs and way of life with the goal that clients pay for the inclusion they need. This expands the intrigue of protection to a more extensive scope of clients, some of whom may then buy protection just because.

- **Risk:** Neural systems can be utilized to perceive extortion designs and lessen fake cases. As indicated by the FBI, non-medical coverage misrepresentation in the US is assessed at over $40 billion every year, which can cost families between $400–700 every year in extra premiums. AI can likewise be utilized to improve Insurance agencies' dangers and actuarial models, which can prompt increasingly beneficial items.
- **Operations:** Chatbots utilizing neural systems can be created to comprehend and answer the majority of client questions over e-mail, talk, and telephone calls. This can save noteworthy time and assets for back up plans, which they can send towards increasingly gainful exercises.

11.4 AREAS WHERE AI CAN HELP THE INSURANCE INDUSTRY

There are numerous instances of how safety net providers around the globe are executing AI to improve their main concern, just as the client experience. There are additionally various new companies that are giving AI answers for safety net providers and clients. I will cover a couple of fascinating cases here.

11.4.1 HEALTH INSURANCE

Health insurance is a crucial component of the Health Industry, and a major revenue source for the Insurance industry. Majorly, AI is serving health insurance in terms of cost efficiency and fraud detection. Cost efficiency persuades the insurance and health industry to work in alliance and develop applications that will foster healthy trends to strategically cut short on preventable expenditures, which are majorly caused due to unhealthy habits and living patterns [11]. The latter utilizes AI using algorithms with structured data to detect cases of fraudulent claims.

In the process, AI brings greater accuracy and technological advancement on the grounds by drawing meaningful patterns and ascertaining information from the profile of each client. In 2018, SwissRe, and Max Bupa Health went into an organization with Indian wellness tech start-up GOQii Health. GOQii utilizes information from wearable gadgets and their own AI-driven "wellbeing motor" to follow wellbeing vitals and give solid living guidance and hazard reports to singular clients [12].

11.4.2 AUTO INSURANCE

AI is taking the Insurance industry by storm. Insurance companies are revising on AI more comprehensive decisions and for providing insurance-related services. Until AI's implementation, providing car Insurance was a tedious and labor-intensive task, but since the introduction of changes by insurtech, the market is changing dynamically. Humans did examine the claims and cases and made judgment calls about the risks. Insurers handled each claim with intensive personal investigations that usually take days (and even months) to complete. Today, AI can bring about transformative change in the digitalized Era by allowing the Insurance Company to provide a highly customized policy wherein they carefully measure the Individual's risk.

As far back as 2017, US safety net provider Liberty Mutual disclosed another designer entryway through its advancement hatchery Solaria Labs. This open API entrance joins open information with restrictive protection information to empower the making of better protection items for clients. One such item was a portable application that permits drivers engaged with mishaps to survey harm to their vehicle, progressively utilizing their cell phone camera. The application would likewise give fixed cost gauges. The AI controlling the application will be prepared to utilize a great many pictures of auto collisions.

Insect Financial, the Chinese fintech firm piece of the Chinese goliath Alibaba Group, discharged programming called Ding Sun Bao to break down auto crash harm and procedure claims. Ding Sun Bao utilizes machine vision, empowering drivers to take photos of their harmed vehicle utilizing their cell phone camera. The application at that point contrasts the photograph, and it's picture database to decide the seriousness of harm, gauge fix costs, and examine the mishap's effect on the driver's future protection premiums. Essentially, Ant Financial cases that the application surveys harms and handles guarantee in six seconds, while human cases agents take around six minutes and 48 seconds.

11.4.3 OPERATIONAL EFFICIENCY

US safety net provider Allstate banded together with the Earley information science (EIS) organization to build up a remote helper called ABIe (articulated 'Abbie') [13]. ABIe was intended to answer basic questions of Allstate's protection specialists who had changed from offering one protection item to another. ABIe utilizes common language preparing to process 25,000 requests for each month.

When exchanging between product offerings, numerous business operators confronted a lofty expectation to absorb information, and Allstate found that its call focuses were immersed with inquiries from their very own business specialists on new items. This prompted long hang tight occasions at call communities for real clients, which brought about conceivably lost business openings. Arrangements like ABIe can conceivably have a tremendous effect.

11.5 FOUR STEPS TO UNLEASH AI ADOPTION IN INSURANCE

With processed data readily available, insurance companies need to bring focus on four pillars of AI adoption that will thoroughly guide them of intelligent AI adoption [14]. Traditionally and decades ago, the world of insurance was manual and was involved with tedious administrative work, forced the executive, and demanded innovative advancement.

The following four pillars have already helped some leading Insurance companies to transform their business model and get AI-ready.

1. **Building a Strategy:** Building a comprehensive and coherent AI strategic plan would help to achieve short and long-term goals. This is done through studying the possible scenarios and trends, aligning on a shared vision, and building a strategic roadmap to achieve the desired future.
2. **Assessing Data and Technology:** Insurance companies have a massive pool of unstructured data, by assessing data fields based on machine learning (ML) and deep learning techniques and categorizing them correctly; Insurance companies actualizes a transformative change allowing them to strengthen products, policies, claims, fraud detection, pricing, and overall system.
3. **Training People:** For AI to get successfully implemented across the organization, Insurance companies should develop or train executives with AI literacy to recognize AI opportunities. AI systems use past decisions to infer rules and learn the unwritten rules and tactic knowledge that are so hard to capture manually.
4. **Building a Governance Model:** With a fast-moving scenario, insurers need to be proactive about designing and implementing governance for AI, including new policies, procedures, and principles to ensure safe, transparent, accountable, ethical, and trustworthy AI.

11.6 HOW CAN AI IN THE INSURANCE INDUSTRY HELP WITH FRAUD DETECTION AND CLAIMS?

From smart chatbots to the array of ML technologies, the expanding potential of AI in Insurance is already reaching its visionary for complete automation. With the game-changing influence of AI in the Insurance Industry, the discomfort around its implementation is fading quickly, and it begins to trust the caliber and numerous opportunities been brought forward by AI and ML in the world of insurance [15].

AI Offers several promising technology-enabled solutions, which include automated claim support, Visual Analytics in claim settlement (damage assessment through pictures and videos), Interactive Chatbots, and Predictive Analytics for proactive measures.

1. **The Uninterrupted Flow of Business Information:** Processed data is becoming an integral part of the Insurance industry; hence how carriers identify, quantify, place, filter, and manage risk is all predicated on the volume and quality of data they acquire during a policy's life cycle. With external data, carriers must focus on securing access to data that enriches and complements their internal data sets. Leveraging AI's data processing capability, an insurer can strategically build a sophisticated environment where information relevant enough to businesses, and customer interaction can flow and get channelize to different departments.
2. **Automated Claim Support:** At the present, Insurance Company, faces the challenge with quick claims and identifying and rejecting the fraudulent claims by customers. Insurance companies particularly are far behind in investing in AI-enabled technology that can not only reduce the tedious manual administrative work but can also enhance the service quality and experience for customers by making the right decisions. The technologically advanced system of AI keeps the agents and employees synchronized, which makes the work a little more transparent and quicker decision-making.
3. **Interactive Power of Chatbots:** With chatbots for faster customer service and with ML technologies that function using the algorithm with structured data, AI enhances the performance of any workplace with its automation process.
4. **Advanced Underwriting:** Insurers leveraging AI, deploy ML, and other models to assess risk and other liabilities. It makes the insurance underwriting more accurate, and also boosts profits by diminishing

risks. It helps to efficiently evaluate risks and outreach potential new customers.

5. **Predictive Analytics for Proactive Measures:** The leadership team's long-term strategic plan will require forming a multi-year transformation that puts effort into operational efficiency, risk management, and technology. As there are macro-level changes in the industry, and there is a line shift towards a "predict and prevent" methodology, carriers will need to rethink and preplan their customer engagement and branding, product designing and pricing and, its core earnings.

11.7 END: BENCHMARKING AI SOLUTIONS IN INSURANCE

The year 2019 has shown substantial growth in Insurance innovations and technology that brought new patterns and value to the industry [16].

Ten remarkable innovations have shaped the development of the Insurance industry:

1. **Augmented Intelligence:** Insurers are augmenting AI to better understand the data patterns of customers and their requirements; also, it competes for the human capacity of Insurance employees during product recommendations, advisory, price quotations based on studying the patterns from customer's profile.
2. **AI-based Smart Automation:** Smart Automation incorporates technological innovation to increase operational efficiency and add value to the customer's final experience. For example, South Korean Kyobo Life Insurance Co. Ltd. has brought about significant development in the Insurance industry through their AI system BARO (best analysis and rapid outcome) to automate underwriting. The system uses NLP (natural language processing) for customer interactions in natural language. Another example is of Religare introduced an AI-based chatbot system in their business model.
3. **Digital Insurance Broker:** Digital transformation has converted the manual and paperwork into digitally automated processes, with the increased use of software and the Internet. A digital Insurance Broker is one which gone through a digital evolution to augment the growth and efficiency of the line of business.
4. **Services Beyond Insurance:** With technological advancement, Insurers have realized to cater the day-to-day needs of the customer for better and richer customer experiences. Medical transport by

drones is an example of 'Beyond Insurance' services to support its customers. As an established insurer of the transport sector, Baloise is a privileged partner in the EU-funded Helicus project, where the organization from the medical and aeronautics sector join hands to test aerial logistics for hospitals in the skies over Belgium.

5. **Blockchain in Reinsurance:** Insurance and reinsurance companies are focusing more on improving the operational efficiency to help sustain and make profitability in today's competitive market, and it is believed DLT, Distributed Ledger Technology, i.e., Blockchain reinsurance technology, could help the same. It is believed reinsurances one of the aspects of Insurance chains that could be helped using Blockchain technology.

   For example, US-based *Aon Benfield and Guy Carpenter* partnered with *Institutes Risk Stream* developed the reinsurance placement process to bring greater efficiency by putting forward brokers and reinsurers on a collaborative platform [17]. Similarly, *CryptoBLK Ltd.*, in collaboration with *The Hong Kong Federation of Insurers (HKFI),* developed Asia's first *Motor Insurance Authentication* platform using blockchain reinsurance technology named motor insurance DLT-based authentication system (MIDAS), to validate motor Insurance policy in real-time.

6. **Unconventional Partnerships:** Insurers' partnerships with Insurtechs, fintech, an integrated technology by insurance, financial services, and other collaborations to improve the delivery to consumers. Some insurers and reinsurers are more active with these partnerships, AXA, Chubb, and QBE are among them. There is a multi-dimensional view over these partnerships in insurtech, as they want to augment their experience with agents and final customers.

7. **Understanding Customers and Delivering Tailored Products:** Meeting with the Customer's customized demand for services, insurers have applied AI-enabled technology to better understand the trends and patterns to enhance the service quality, and hence, customer experience. For example, UK-based *Bought by Many-Pet Insurance* helped people to insure their pets.

8. **On-demand Insurance Models:** Technological innovations have significantly changed over several years, and it has even brought drastic changes in consumer expectations, hence transforming the way business has been conducted. On-demand Insurance models allow the customer to choose cost-effective and convenient pricing

and coverage based on their needs. An attractive user interface, one-click purchase, and customized coverage have contributed to the appeal for on-demand insurance.

9. **Risk Intelligence:** Insurers leveraging AI, deploy ML, and other models to assess risk and other liabilities. It makes the insurance underwriting more accurate, and also boosts profits by diminishing risks; and it helps to efficiently evaluate risks and outreach potential new customers.
10. **Customer Educations:** Educating customers about prospective risks isn't sufficient; mixing this information with available products' prices and benefits is a must.

11.8 AI-BASED TECHNOLOGY AND TRENDS RENEWING INSURANCE BUSINESS MODEL

AI-underlying technology is already being used in today's businesses at a global pace. Four core technological trends by AI will shape the insurance industry over the next decade:

1. **Extracting Data from Connected Devices:** The penetration of consumer connected devices (such as smartwatches, fitness tacks, car gadgets) will continue to reap or extract data which will allow users to understand their client deeply, resulting in a new product line, more personalized pricing and increasingly real-time service delivery.
2. **Advancement of Physical Robotics:** The field of robotics has seen rapid growth and achievements, and this innovation will continue to serve the globe for over a few more decades to come. By 2025, Construction 3D Printing (c3Dp) will be prevalent, with various technologies that use 3D printing to fabricate buildings. Prospective benefits of these technology-driven buildings include faster construction, low labor costs, greater accuracy, and integrity with judicious use of available resources.
3. **Development through Cognitive Models:** NN and other deep learning tools based on human's cognitive ability to handle large and complex data streams will be generated and tied to an individual's behavior and activities to study the patterns of human behavior.

   With the increased advancement and commercialization of these technologies, carriers will be constantly learning and forming models to be adopted by the world around them, while responding to risk management and behaviors in real-time.

11.9 HOW INSURERS CAN PREPARE FOR ACCELERATING CHANGES?

The rapid evolution of the Insurance industry with the extensive adoption and incorporation of automated technology, deep learning, and data ecosystem, providing new heights to the standard of the line of business in insurance. The following pointers will fetch insurers with sufficient data to accelerate with the required changes in the industry:

1. **To Renovate on AI-based Technologies and Trends:** Although AI-related trends in the industry are tech-focused, also customer-experience team (along with the IT team) should invest time and resources to build a deeper understanding of the workings of these AI-based technologies.
2. **Develop and Execute Coherent Strategic Plan:** The leadership team's long-term strategic plan will require forming a multi-year transformation that puts the effort into operational efficiency, risk management, and technology. As there are macro-level changes in the industry, and there is a line shift towards a "predict and prevent" methodology, carriers will need to rethink and preplan their customer engagement and branding, product designing and pricing and, its core earnings.
3. **Identify and Implement Comprehensive Data Strategy:** Processed data is becoming an integral part of the Insurance industry; hence how carriers identify, quantify, place, filter, and manage risk is all predicated on the volume and quality of data they acquire during a policy's life cycle. With external data, carriers must focus on securing access to data that enriches and complements their internal data sets.

11.10 THE STATE OF INSURANCE IN 2030

With continuous evolution and a new wave of deep learning techniques, AI has the potential to develop and advance insurance to a state of being "Preventive" and detecting the fraud claims and involving inaccurate underwriting, hence transforming each aspect of the industrial model. An anticipated pace of the change will advance as brokers, underwriters, insurers, financial intermediaries become more adaptive to the change, using advanced technologies for improved decision making and productivity, lower costs, reduced fraud claims, and better service quality.

KEYWORDS

- **artificial intelligence**
- **fraud detection**
- **Hong Kong Federation of Insurers**
- **insurance**
- **Insutech**
- **neural networks**

REFERENCES

1. Artigas, J. C., (2010). *Linking Global Money Supply to Gold and Future Inflation*. World Gold Council, Gold: Report, WGC-US-INVE003, 2. New York: World Gold Trust Services, LLC.
2. BIS, (2015). *Global Liquidity: Selected Indicators*. Bank for International Settlements, Statistics. http://www.bis.org/statistics/gli/gli_feb15.pdf (accessed on 1 December 2020).
3. CGAP & WB, (2010). *Financial Access 2010: The State of Financial Inclusion through the Crisis* (pp. 5–11). Consultative Group to Assist the Poor, The World Bank Group. The United States of America.
4. Davies, J. C., (1962). Toward a theory of revolution. *American Sociological Review, 27*(1), 5. California Institute of Technology.
5. Dillow, C., (2015). *What Obama's New Defense Budget Means for the Aerospace Industry*. Fortune Magazine, Tech, Aerospace.
6. Egan, M., (2014). *Six Companies behind the Drone Revolution*. CNNMoney. CNN News Channel.
7. Federal Aviation Administration (FAA). (2016). *Regulations and Policies*. Commercial Space Transportation, Final Rules. https://www.faa.gov/space/legislation_regulation_guidance/ (accessed on 8 March 2021).
8. Garver, R., (2015). *Yikes! Putin Will Soon Have Superhuman Robo-Soldiers*. CNBC News Channel.
9. Hern, A., (2015). *Samsung Rejects Concern Over 'Orwellian' Privacy Policy*. The Guardian Newspaper, Tech.
10. Karlan, D., & Morduch, J., (2010). Access to finance. In: Dani, R., & Mark, R., (eds.), *Handbook of Development Economics* (Vol. 5, pp. 4703–4784). Chapter 2.

11. Keller, G., (1983). *Academic Strategy: The Management Revolution in American Higher Education* (p. 211). The John Hopkins University Press, Education.
12. Lolli, F. M. A., (2013). *Semantics and Pragmatics in Actual Software Applications and Web Search Engines: Exploring Innovations*. CoRR abs/1307.0087.
13. Manyika, J., Chui, M., Bughin, J., Dobbs, R., Bisson, P., & Marrs, A., (2013). *Disruptive Technologies: Advances That will Transform Life, Business, and the Global Economy* (Vol. 180). San Francisco, CA: McKinsey Global Institute.
14. Öztürk, B., & Mrkaic, M., (2014). *SMEs' Access to Finance in the Euro Area: What Helps or Hampers?* IMF Working Paper, WP/14/78. International Monetary Fund, European Department.
15. Pau, L. F., (1986). *Artificial Intelligence in Economics and Management*. Elsevier Science Inc.
16. Taylor, J. G., (1993). *The Promise of Neural Networks*. Springer-Verlag New York.
17. Wainwright, O., (2015). *Let's All Move to Mars! The Space Architects Shaping our Future*. The Guardian Newspaper.

CHAPTER 12

IMPACT OF ARTIFICIAL INTELLIGENCE IN CHANGING TRENDS OF MARKETING

SEEMA SAHAI and RICHA GOEL

Amity University, Noida Campus, Uttar Pradesh, India,
E-mail: ssahai@amity.edu (S. Sahai)

ABSTRACT

This chapter dives deep into the concepts of artificial intelligence (AI) and presents extensively studied facts on ongoing trends in marketing. In this fast-changing digital era, digital marketing is geared up to insert the sense of being effective and efficient than using traditional marketing concepts. The use of AI has become one such paradigm in the era of digital marketing. Light is thrown at ongoing trends in marketing, as the topic suggests studying the impact of AI and also knowing its role in value co-creation, resource integration, and service for service exchange. While covering the issues of these areas, the researcher has come across more prevalent problematic areas, which are known and cannot be avoided, like employment and ethical aspects.

Research is primarily based though there is an input of expert advice, in which thorough effort and quality information are provided. The research has just tried making it clear that whether AI is building up excitement or nervousness in upcoming years of marketing. Many types of research related to AI are either purely scientific-based or talk about specific products with innovative inbuilt functions or, at the most, have come up with collective positive and negative perceptions. Moreover, a country like India has got a huge economy with young aspiring software engineers and marketers, so to explore and study the above-mentioned aspects for this country can bring in new theories, perceptions, employment opportunities, innovative ideas, and so on. This research can add extra bling to the references for the upcoming research projects.

12.1 INTRODUCTION

To start with our extensive research, let's first understand the concept of marketing. One of the renowned faces of marketing, Philip Kotler, has defined marketing in simple words, i.e., marketing is all about the satisfaction of customer's needs and wants through a process of exchange between buyers and sellers. Marketing as a concept is built on the foundation of three golden words that are needs, wants, and demands. Marketing has evolved from traditional practices to modern practices. In earlier times, marketing was product based as more and more number of customers grew it was production-based, and further, it got shifted to mere sales soon organizations realized that customer's need is necessary to function in the market and there emerged the concept of marketing and to have goodwill and say in the market and the society as a whole it came up with a modern philosophy of societal marketing. Now to run the concept of marketing to its successful terms, the 4P's have a vital role to play in this, i.e., product, place, price, and promotion. To apply the concept of marketing, you need to have a product or service which is to be marketed, and the place is all about the distribution patterns and processes, price of the product or service (which is extensively used to lure customers), and the most important concept and one of the strong pillars of marketing-promotion, this is a golden tool used by all organizations, it is inclusive of all communication types in marketing. Therefore, these are some of the most commonly talked about concepts of marketing.

12.2 DIGITAL MARKETING AND ARTIFICIAL INTELLIGENCE (AI)

Digital marketing is the marketing of products/services using digitally efficient technologies, generally using the Internet, which also includes cell phones, display advt., and any other form of the digital medium. Digital Marketing has become a medium for brand awareness. This very brand awareness helps in the following:

There is ease of access:

1. It gives organizations a competitive advantage.

2. It is effective in countries where they are efficient in avoiding uncertainty related to markets and the economy.

Digital marketing is solely dependent on evolving, innovative, and fast-changing technology; with this, digital marketing strategies and developments have to get altered on a timely basis. Hence segmentation and influencer marketing is given more importance and are extensively focused upon in digital marketing. With recent trends, new developments are seen like: online behavioral advertising, collaborated environment, remarketing, game advertising, and data-driven advertising. An interesting fact is that marketers have democratized the promotional landscape by using digital tools. One strategy used for being effective in the area of digital marketing is content marketing, and to build long-term relationships with customers, the marketers have found another important strategy for digital marketing that is marketing through e-mail. Digital marketing uses many channels to operate; the most extensively used channels are Affiliate Marketing, display advertising, search engine marketing, e-mail, and social media marketing, online PR, etc.

It is very evident how strategies are made and implemented. In this era of highly efficient and effective markets, it is important to have proper digital marketing strategies. One such important step is planning when planning is done rightly, mostly all steps follow on the right track. Planning involves three stages, i.e., opportunity, strategy, and action. Marketing activities like co-creation of value, integration of resources, and service for exchange of service are also gaining a lot of importance through the use of artificial intelligence (AI). In developed nations, it's quite successful, while in developing nations, it is still trying to keep up with the standards. However, online shopping websites in India like Myntra, Jabong, etc., have tried their level best in creating value co-creation through digital marketing and are indirectly using AI technology, which applies for resource integration and service for service exchange. However, many companies involved in the supply chain or other activities rely on physical or manual work, though it has some positives of human employment.

As technology is innovative and fast-changing, AI, also known as machine intelligence (MI), is gaining a lot of importance in digital marketing. AI commonly used the word that consists of different approaches and technologies that are set up in such a manner to "think" like human beings. It is already inbuilt in multiple technologies like voice recognition and chatbots.

In this chapter, we will see and explain why AI is about to become an integral part of almost every aspect of the fast-changing trends of the digital marketing world. There are many benefits of introducing AI in business organizations; three such known benefits are:

1. AI, when introduced, leads to lesser mistakes;
2. It helps organizations with daily applications;
3. Forms and provides a base for digital help.

AI in digital marketing levels up the user experience, predict buyer behavior, gains on ROI, Hassle-free, and smarter search sessions, getting hold and targeting of the right audience, easy prediction of sales forecast, and improvement in the advertising sector.

Point to be noted, i.e., AI can work well with digital marketing; therefore, it also has its target audience:

1. New startups/companies, etc., and digitally advanced marketers;
2. Content creators involved with content creation and curation;
3. Generation leading newbies, e-mail marketers, and experts, etc., who are interested to learn and associate with AI.

AI has got a stronghold in the production line, which will first affect the employment opportunities at the lower levels and will also affect supply chain marketing (SCM). The sub-fields of AI like artificial neural networks (ANN), machine learning (MI), Expert systems, Natural language, and computer vision have made it's self an important part in terms of their usefulness for improving the effectiveness and efficiency in marketing. Product's like Apple's Siri, Google's Alexa speakers, and self-driving cars, Facebook's image recognition are popular examples of AI. The marketers smartly marketed the AI technology of these products. Many online brands like Myntra, Jabong. Amazon, Flipkart, and others have also extensively used innovative techniques of AI to capture the markets. As our topic suggests, one thing we have noted is these companies do change or keep on updating their products concerning the ongoing trends in the market with the help of AI. Therefore, we can identify that with new market trends, the marketers have to bring in new changes in their AI.

Each of the subfields of AI has greatly created an impact on marketing trends. These marketing trends are known from the consumer's buying behavior, needs, wants, and demands. The fact is people are using AI but don't know its aspects in depth. Some are excited about its use, and some are nervous. By a few researchers, it predicted that many employees might lose their jobs by 2030 by a strong and huge intervention of AI, with these ethical calamities are sure to rise.

12.3 LITERATURE REVIEW

Venugopal and Baets [1] discuss ANN and tell that how earlier it was used in business and present-day then and in future how it is going to benefit the businesses. The researchers have brought in the pros and cons, similarities, and differences between ANNs and statistical methods as they were confusing terms for the general employees. Market researchers and managers who are inclined to some support system for decision-making can refer to this chapter to get an insight into the concept. Mulvenna and Buchner [2] describe the combing of datamining techniques on internet data to obtain marketing intelligence. In general, views are materialized, and different types of online analytical web usage mining methods have been described.

Culotta, Ravi, and Cutler [3] predicted the demographics by the fact that people on Twitter follow whom and why. They have also used a supervised learning approach in which later they discovered that marketers have the upper hand in social networking sites (SNSs).

Cheung, James, Law, and Tsui [4] throw light on the issues faced by systems like content-based and collaborative systems. It also extensively studies the ML algorithms like the latent class model and support vector machine, according to the researchers, which will help in solving problems. Swaminathan, Smith, and Sadeh [5] briefs about supply chain modeling framework which was designed to overcome the problem of pragmatic approach which was hampering the effort and time needed to make models with efficient fidelity to the interest of the supply chain. This chapter also discusses the software components that represent types of middlemen in the supply chain models. They have done this by analyzing various supply chains. Li and Shuliang [6] discussed the topic. In addition, the architecture of the software of hybrid systems is discussed. Summarization of empirical fieldwork on evaluating the hybrid system is done in which the findings show that hybrid systems are beneficial for marketing strategies.

Venugopal and Beats [7]—like neural networks (NN) have become important for the intelligence of the business; the researchers have aimed to explain the concept in the simplest of terms by describing the basics of NN and also have presented a review of selected work done in the field of ANN in marketing sciences. Rao, Nahm, Shi, Deng, and Syamil [8] suggest the researchers review and categorize them later. The articles are grouped as per new product development stages and seven core elements (competencies and abilities). Future research areas are directed.

Chiu [9] described the developed and tested genetic algorithm—case-based reasoning (CBR) system that will predict customer purchasing behavior. This system was tested and developed by an insurance direct marketing co. in Taiwan. Francisco, Lopez, and Casillas [10] cover a journey from the 1970s to 2013 of the application of AI with a focus, especially on industrial marketing. They have provided conclusions and a few future insights. They have also presented a set of papers on this very special issue. Wierenga and Berend's [11] study instructs marketers to take help and start relying upon intelligent systems. However, at times strategic issues like the development of a product, innovativeness, and decisions for long-term goals concerning targeting, positioning, and segmenting, etc., do need the support of AI systems. They have explained that a marketer's intuition and emotions can prove fatal for markets and decision-making.

Kozinets [12] created ethnography as a technique for online research marketing to get an idea about consumers. According to the author, the technique is simple, fast, and not that expensive. The author shows a netnography of an online coffee newsgroup and discusses marketing applications. Walle [14] describes the qualitative aspect of intelligence and marketing when combined. Many researchers go off the track when they talk about AI, and they start focusing on the technical areas which are unknown to the general public that includes marketers too. This book is a good guiding and motivational source for marketers.

Changchien and Lin [15] focuses on designing the architecture of CBR and also a method to help in sharing and to retrieve back the cases of important concern to the marketers. The CBR was implemented in a telecommunication co. to prove this research. Baesans and Verstraeten, Poel, Peterson, Kenhove, and Vanthiens [16] study are that they have compared and evaluated many Bayesian network classifiers with statistical technique and many other AI techniques with a purpose to classify the customers. They also extensively study the very same networking classifiers, customers' buying logic, in particular, and their psychology.

Fish, Kelly, Bames, Milan, and Aiken [17] justifies that by the use of these networks, the marketers can segment their territories more easily with correct data, which will further improve sales and other promotional activities. The authors go in-depth to the potential of ANNs, which will help marketers with the segmentation difficulties by compared analysis of discriminant analysis with logistic regression. McDowell, Forrest, and Bogdan [18] argue that instead of exploiting the customer's desires and weaknesses, why not focus on the user's actual needs and market them. The overall chapter talks about their perception on long-term implications, they have also explored the role of trust issues within customers concerning the adoption of AI agents, and they have also brought up the issue of the security needs for agents and the ethical aspects to the access to these agents.

Syam and Sharma [19] have perceived that in the next 10 years, the world will witness an aggressive industrial revolution with the support of digitalization, information, and communication technologies. MI, robotics, and AI, hence resulting in a huge impact on personal selling and other sales management. The researchers have studied the impact that is related to the steps of the process of selling. Siau and Yang [20] talks about employment issues that the world will see and suffer due to the encroachment of AI technology. The research is a longitudinal case study that analyzes the fast-changing nature of the sales and marketing aspects. Gursay [21] opinionates the experts intending to give an overview of the development of hospitality marketing. This is because it is seen that the focus on research on the same has shifted. There is less focus on the development of products and more on consumer's evaluation of the products, which also influences their decision-making. Michaels [22] talks about balancing out the use of AI technology and employment opportunities for the welfare of the economy. They have extensively discussed the impact of AI in various fields, particularly in marketing. They have discussed the positives for marketers by proving AI as a tool and not as a competitor for employment.

12.4 RESEARCH PROBLEM IDENTIFIED

By investigating and thorough reading of these many, it was noticed that AI has made progress but has to attract the acceptance of many corporates and individuals in business organizations. Moreover, marketing trends are also changing at a fast pace, and to cope with these Artificial Intelligent technologies and corporates are finding it difficult to operate, they are keeping old methods of working handy to overcome failure and lateness. Many marketers are unaware of the importance of AI in value co-creation, resource integration, and service for service exchange. While covering the issues of these areas, the researcher has come across more prevalent problematic areas that are known and cannot be avoided, like employment and ethical aspects.

Thus a survey of 60 respondents which are new owners of successful startups, managers from marketing fields, and students further pursuing marketing, was done.

1. **Adoption of AI in the Marketing Sector:** The survey was taken amongst 60 respondents, amongst the age group of 18–62 years. On being asked about their preference for the adoption of AI in the marketing sector, amongst them, 33.9% of them strongly agreed, 42.4% of them agreed. Only 20.3% of them were neutral, and 1.7% of the people disagreed. Looking at the chart, we can conclude that most of them are in favor of adopting AI in the marketing sector (Figure 12.1).

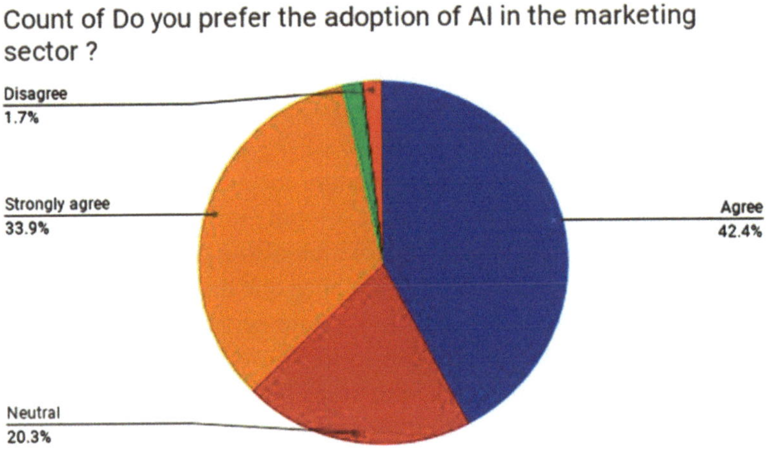

FIGURE 12.1 Graph of adoption of AI in the marketing sector.

2. **AI Helping Marketers Over the Years:** As the technology has been evolving tremendously, people's take on AI catering to marketers over the years had strong views, 37.3% of them strongly agreed while 35.6% just agreed. 18.6% of the people were neutral, and surprisingly 3.4% strongly disagreed with the statement, and 5.1% just disagreed. Here the conclusion lies that the majority of them believe that AI has helped marketers over the years (Figure 12.2).
3. **AI Promising for the Marketing Sector:** As said earlier, technology is taking a fast pace, many questions are raised in the future. 39% of the people strongly agreed, while 37.3% just agreed on AI being promising for the marketing sector. 18.6% of the people were neutral. 3.4% of them disagreed. Hence, the majority of them believe that AI would emerge as a promising aspect of the marketing sector (Figure 12.3).

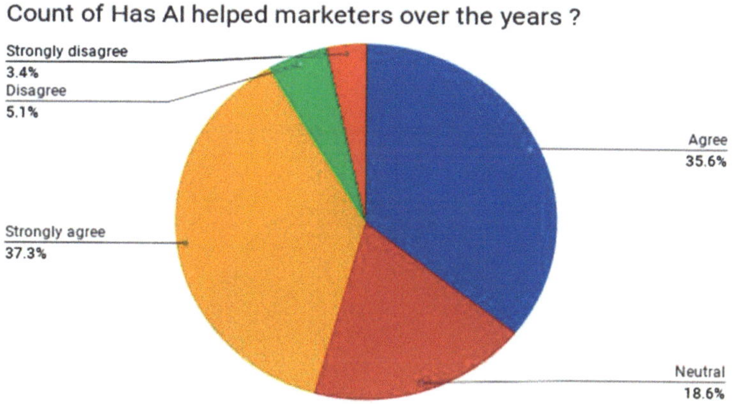

FIGURE 12.2 Graph of AI helping marketers.

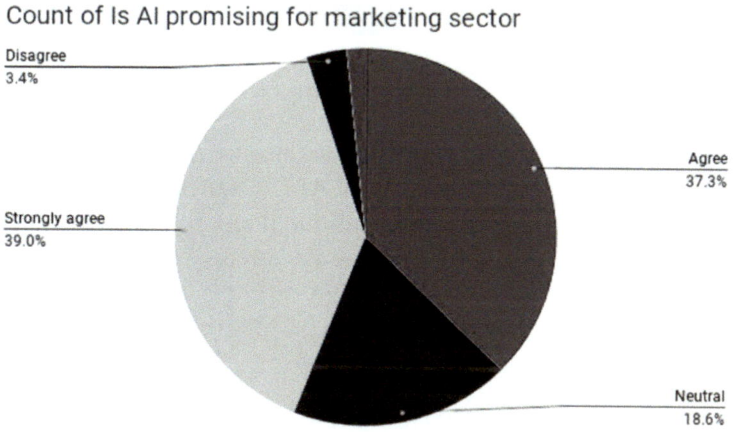

FIGURE 12.3 Graph of AI promising for the marketing sector.

4. **People's Take on AI Technology Helping Marketers in Making Related Decisions:** Machines and human minds always conflict with being better off. However, the point that is to be noted that machines can never work and make decisions with emotions. On being asked that by adopting AI technology can it help marketers in making market-related decisions, amongst them, 32.2% agreed while 3.4% strongly agreed with the statement. 35.6% of them were neutral, 11.9% of them strongly disagreed, while 16.9% just disagreed with the same. Here the conclusion lies that less than 50% of them are supporting the statement, which indicates that people are still not

confident about AI technology in the usage of making market-related decisions (Figure 12.4).

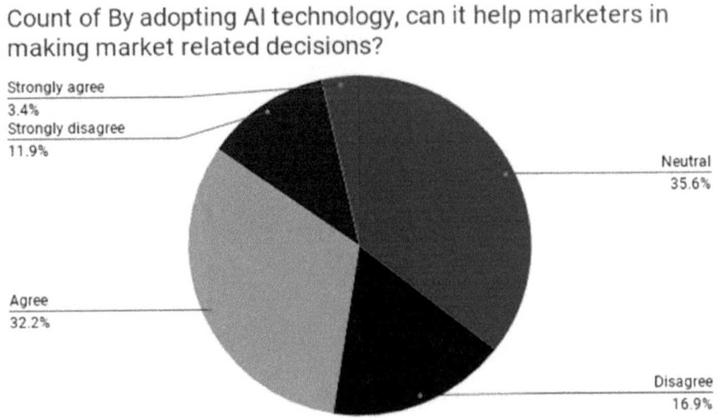

FIGURE 12.4 Graph of adoption of AI.

5. **Transforming of Jobs and Industries in the Marketing Sector:** On this, many people believe that AI has transformed jobs and industries over the years and will continue doing so. 20.3% of them were neutral. Only 3.4% of the people disagree with the statement while the other 3.8% of them surprisingly strongly disagree with the same (Figure 12.5).

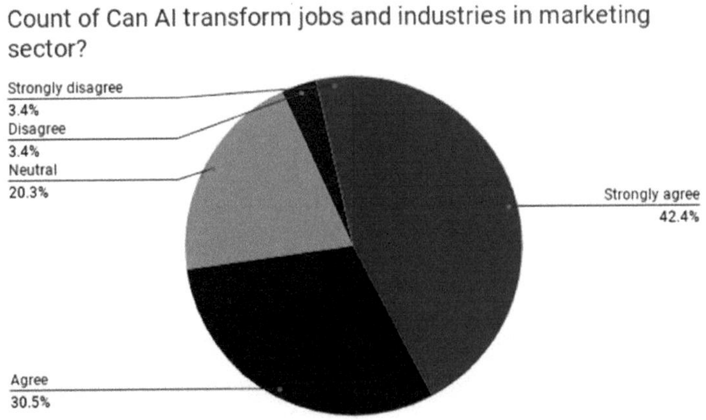

FIGURE 12.5 Graph of AI transforming jobs in the marketing sector.

6. **Artificial Intelligence Catering to the Development of Human-Level Intelligence, Creativity, and Motivation, which is Required in Marketing Activities:** At present, we all rely a lot on machines for every aspect of our lives, considering AI's role in helping to develop human-level intelligence, creativity, and motivation, 50.8% of the people were neutral, while surprisingly 10.2% of them strongly agreed and 23.7% of them just agreed. However, they were some who strongly disagreed with 1.7% and just disagreed 13.7%. Here we can conclude that people were neutral with the statement and are also able to balance with the machine and natural life (Figure 12.6).

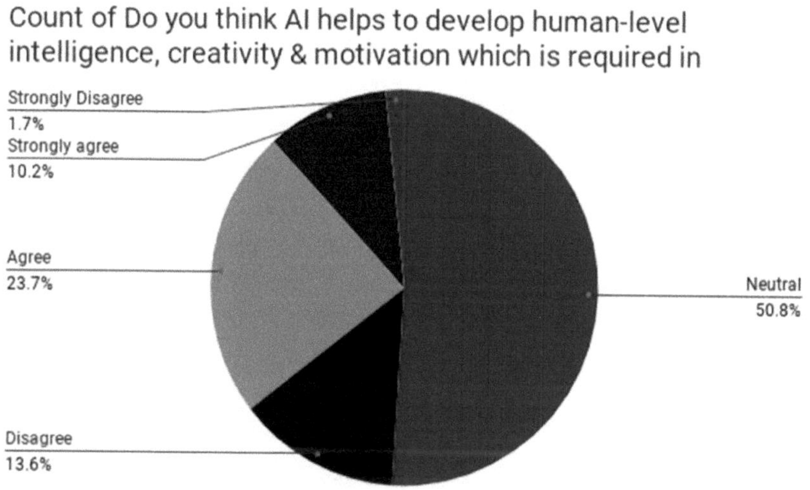

FIGURE 12.6 Graph of AI develops human-level intelligence.

7. **Take on Artificial Intelligence being Limited to Production:** Machines can never make a mistake if programmed accordingly, and this best suits for production line or for anything which requires continuous and tireless effort. Around 40.7% of the people agree with the statement, and 6.8% just agree. While 28.8% of them are neutral. 20.3% of the people disagree with the same, and 3.4% of the people strongly disagree. Here People are divided greatly on their opinions, but if we go by the numbers, we can conclude that less than 50% of the people believe that AI should be limited to production (Figure 12.7).

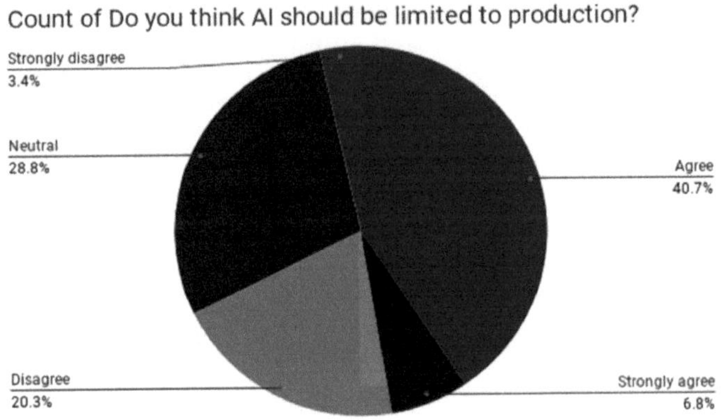

FIGURE 12.7 Graph of AI limited to production.

8. **Feasibility of Artificial Intelligence:** Often, questions are raised that in this fast-changing technological phase, AI will be faithful to long-term goals. On being asked about the feasibility aspect of AI for long-term goals, 50.8% of them agreed with the statement, and 16.9% of them strongly agreed. 28.8% of them were neutral. At the same time, a very small percentage of 3.4% disagreed with the same. Therefore, the majority of them believe that AI is feasible for long-term goals (Figure 12.8).

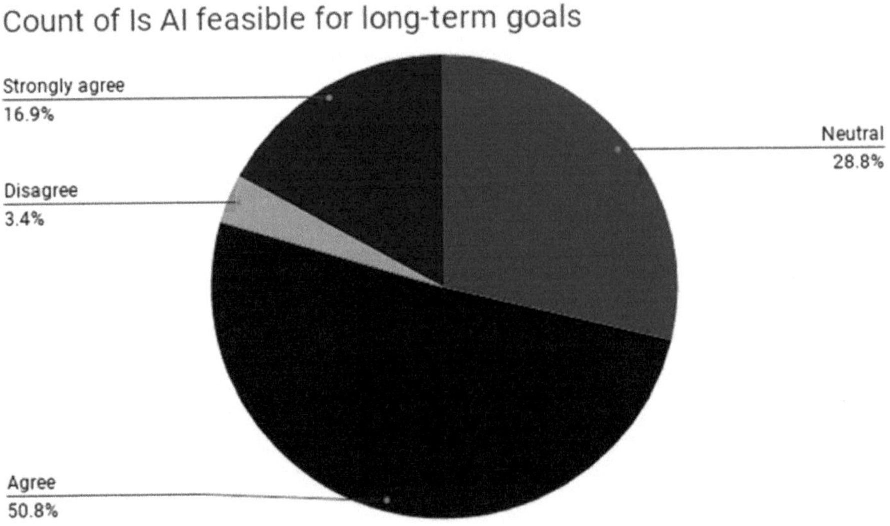

FIGURE 12.8 Graph of AI for long term goals.

9. **Value Co-Creation in Artificial Intelligence:** Value co-creation being an important aspect has proved this statement as 49.2% of them agree, and 6.8% of them strongly agree, and 42.4% of them were neutral. While just 1.7% of them strongly disagreed (Figure 12.9).

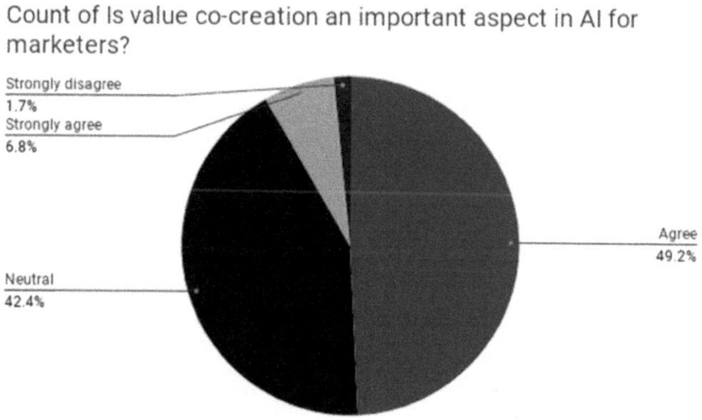

FIGURE 12.9 Graph for value co-creation in AI.

10. **Supporting of Fast-Changing or Inconsistent Trends in Artificial Intelligence:** Inconsistency is disliked by all the organizations, but marketing is known for that, but the whole of that changes if thought on the lines of AI.8.5% of the people strongly agreed on supporting the fast-changing or inconsistent trends in AI and 27.1% of them agreed. While 28.8% were neutral. 32.2% disagreed with the statement, and 3.4% strongly disagreed. The conclusion is people are having mixed views on this statement (Figure 12.10).

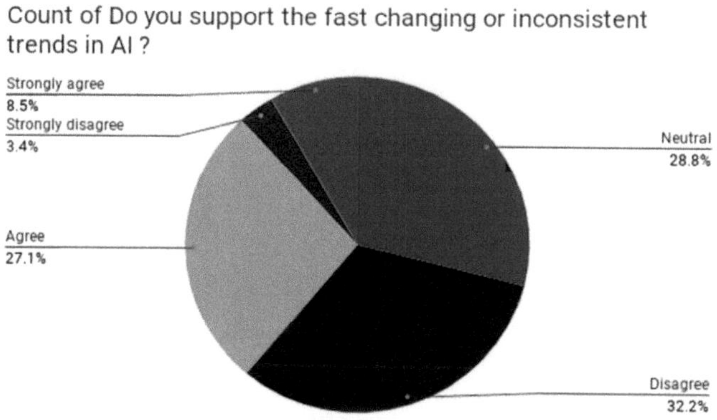

FIGURE 12.10 Graph of fast-changing or inconsistent trends in AI.

11. **Virtual Interaction while Purchasing a Product or Availing a Service:** In the era of emerging online shopping concept, we asked if as an individual are they comfortable with the virtual interaction while purchasing a product or availing a service, 33.9% of them agreed, and 5.1% of them strongly agreed. 39.0% were neutral. 18.6% of them disagreed, while 3.4% strongly disagreed (Figure 12.11).

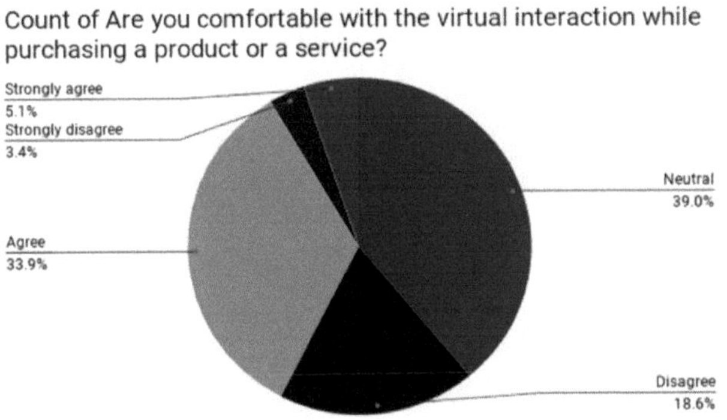

FIGURE 12.11 Graph of comfort with virtual interaction while purchasing.

12. **Artificial Intelligence Creating a Positive Impact:** For this statement, 35.6% of them agreed, and 10.2% strongly agreed. 44.1% were neutral. 8.7% of them disagreed, while 1.7% of the people strongly disagreed with the statement. Therefore, many believe that AI is creating a positive impact on them (Figure 12.12).

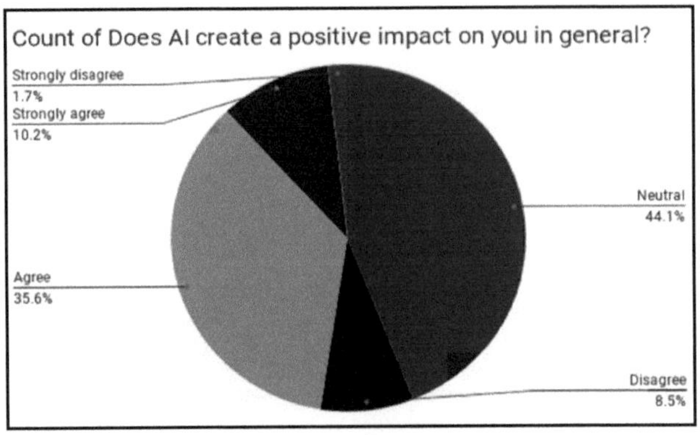

FIGURE 12.12 Graph of the positive impact of AI.

13. **Ethicality of Artificial Intelligence:** Often, ethical issues have been raised on the usage of AI. On being asked the same, 59.3% of them were neutral. 25.4% of them agreed with the statement, and 6.8% of them strongly agreed. While 3.4% of people strongly disagreed and 5.1% disagreed. Therefore, many people consider the usage of AI to be ethical (Figure 12.13).

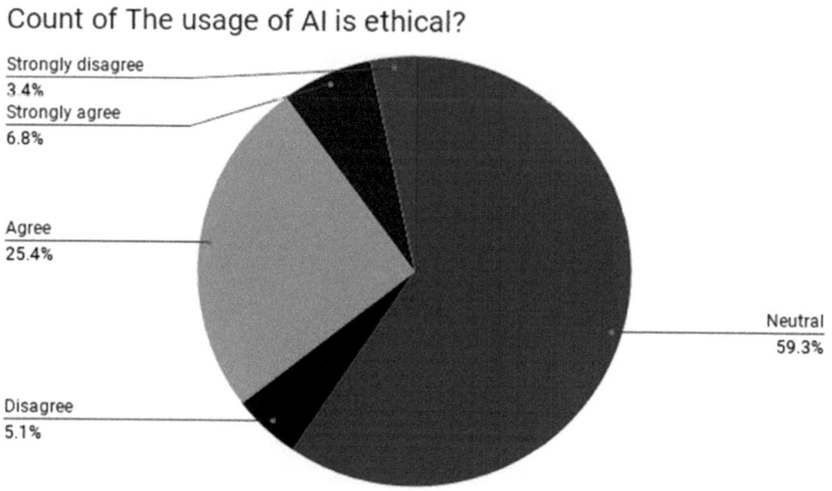

FIGURE 12.13 Graph of the ethical usage of AI.

14. **Awareness of the Changing Trends in AI Technology:** It is important to know the ongoing trends in the market. On being asked about the awareness of the changing trends in AI technology, 47.5% of the people agreed, and 18.6% of the people strongly agreed while 25.4% of them were neutral, and 8.5% of them disagreed. Therefore, the majority of them are aware of the changing trends in artificial technology (Figure 12.14).
15. **Reluctance to Adopt Artificial Intelligence:** As the main question arises on how many of them are willing to adopt the whole of artificial technology, this survey has questioned professionals on the same; 47.5% were neutral, and 39.0% have agreed with the statement while 10.2% of the have strongly agreed, and 3.4% disagreed. Therefore, many of the existing and emerging professionals are willing to adopt AI into their core business activities (Figure 12.15).

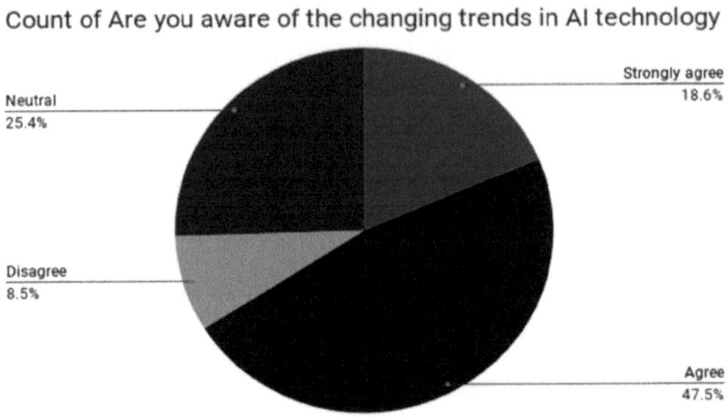

FIGURE 12.14 Graph of awareness of changing trends in AI.

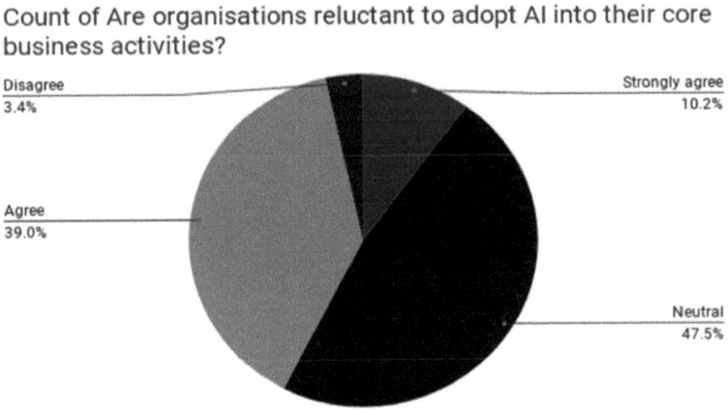

FIGURE 12.15 Graph of reluctance in adopting AI.

12.5 FINDINGS AND CONCLUSION

By the interpreted data which was collected, we found many aspects or perceptions of people on AI. First, the age group of 18–62 years had participated in the survey, amongst which youngsters from 18–33 years had still many positive perceptions about AI while 34–62 years' age group were balanced and some were not so positive about the concept.

Most of them are in favor of adopting AI in the marketing sector and believe that AI has helped marketers over the years. There is also a strong belief that AI would emerge as a promising aspect of the marketing sector. But people are still not confident about the AI technology in the usage of making market-related decisions and are not that convinced with the fact that

AI helps in motivating and being creative in market-related activities. People have a strong take that in the coming years' AI is going to revamp the work culture and is going to greatly affect the jobs. Following this, many think that it should be limited to production alone. The majority of them believe that AI is feasible for long-term goals; following this, they do not support the fast-changing or inconsistent trends in AI. In this, we can assume that people and organizations take time to adjust to one system, and in this fast-changing technological era, it becomes difficult to cope up with the trends.

Surprisingly many people were aware of the importance of AI in value co-creation. On being asked about the online shopping experience, reactions were mixed. In this era of online shopping, many of them are comfortable with it, and some feel forced to go with the trend. Organizations will have to work on their service for the service exchange aspect and make a better experience for customer satisfaction. Overall, people are comfortable with the impact of AI on them and consider it to be positive. We can assume that they have concerns over the security of their personal and public data but still consider it to be ethical, which itself is a big thought. The irony is many don't even know that they are using products that are part of AI.

On being asked about whether they would like to adopt AI into the organization's core activities, many of them agreed upon the statement while many of them were neutral. From this, we can assume that many are excited about the implementation while many of them are nervous as this might hamper their career and may give rise to employment issues.

The final take is, many experts who have worked in the marketing sector for almost 30 years believe and have their logical point that in a country like India, the concept of AI is at an ascent stage. A lot is yet to be experienced. As stated in the survey, people are fine with its use in production as manufacturing is all-physical and involves heavy-duty, whereas marketing is part emotional. Now it's emotional because of the diverse culture we live in. For, e.g., Google's Alexa is creating storms in the Indian market, but it can be purchased only by affluent families, and they don't know its implications. People are buying because of their human emotions. Google's Alexa runs on command, but sentiments with a 24-hour radio station, which is operated and hosted by a human being, will always have the upper hand. There are many more examples, but what we see is more machines and less human interaction, which will lead to an emotionless economy. It's true that markets run on a consumer's needs, wants, and desires, and these things arrive from emotions, and when there are no emotions, the question lies how the economy is going to work? This thus creates excitement and nervousness with the whole concept of AI. The only recommendation that can give is balancing between machines and human emotions, and in any case, there should be the slight upper hand of human emotions. Therefore, AI is yet to fathom human sentiments to a greater extent in a country like India.

KEYWORDS

- **artificial intelligence**
- **artificial neural networks**
- **digital marketing**
- **e-commerce**
- **machine intelligence**
- **search engine**

REFERENCES

1. Venugopal, V., & Baets, W., (1994). Neural networks and statistical techniques in marketing research: A conceptual comparison. *Marketing Intelligence and Planning, General Review, 12*(7). MCBUP Pub. Ltd.
2. Mulvenna, M., & Buchner, A. G., (1998). *Discovering Internet Marketing Intelligence through Online Analytical Web Usage Mining.* ACM Sigmond Record, Source DBL. doi: 10.11.45/306101.306124.
3. Culotta, A., & Ravi, N. K., & Cutler, J., (2015). *Predicting the Demographics of Twitter Users from Website Traffic Data* (Vol. 1). Publisher- AI Access Foundation (Marketing).
4. Cheung, James, T., & Law, M. H., & Tsui, (2003). *Mining Customer Product Ratings for Personalized Marketing, 35*(2). Elsevier Publications.
5. Swaminathan, J. M., & Smith, S. F., & Sadeh, N. M., (1998). *Modeling Supply Chain Dynamics: A Multi-Agent Approach* (Vol. 29, No. 3). Published-2007, Wiley Online Library, Decision Sciences.
6. Li, S., (2000). *The Development of a Hybrid Intelligent System for Developing Marketing Strategy, 27*(4). Elsevier Publications (Marketing).
7. Venugopal, V., & Beats, W., (1994). Neural networks and their applications in marketing management. *Journal of System Management, 45*(9). Cleveland.
8. Rao, S. S., Nahm, A., Shi, Z., Deng, X., & Syamil, A., (1999). AI and expert systems applications in new product development: A survey. *Journal of Intelligent Manufacturing, 10*(3/4).
9. Chiu, C., (2002). A case-based customer classification approach for direct marketing. *Expert Systems with Applications, 22*, 163–168. Elsevier Sciences, Pergamon.

10. Francisco, J., Lopez, M., & Casillas, J., (2013). AI-based systems applied in industrial marketing: A historical overview, current, and future insights. *Industrial Mktg. Mgmt., 42*(4), 489–495. Elsevier Pub.
11. Wierenga, B., (2010). Marketing and AI: Great opportunities, reluctant partners. *Marketing Intelligence Systems Using Soft Computing, 258*. Springer.
12. Kozinets, R. V., (2002). The field behind the screen: Using netnography for marketing research in online communities. *Journal of Marketing Research, 39*(1), 61–72.
13. Crone, S. F., Lessmann, S., & Stahlbock, R., (2006). The Impact of preprocessing on data mining: An evaluation of classifier sensitivity in direct marketing. *European Journal of Operational Research, 173*(3), 781–800. Elsevier Pub.
14. Walle, A. H., (2001). *Qualitative Research in Intelligence and Marketing (The New Strategic Convergence)*. Quorum Books Westport Connecticut, London.
15. Changchien, S. W., & Lin, M. C., (2005). Design and implementation of a case-based reasoning system for marketing plans. *Export Systems with Applications, 28*, 45–53. Elsevier Pub.
16. Baesans, B., Verstraeten, G., & Poel, D., Peterson, M. E., Kenhove, P. V., & Vanthiens. J., (2004). Bayesian network classifiers for identifying the scope of the customer lifecycle of life-long customers. *European Journal of Operational Research, 156*, 508–523.
17. Fish, K. E., & Bames, J. H., & Milan, W. A. A., (1995). Artificial neural networks: A new methodology for industrial market segmentation. *Industrial Marketing Management, 24*(5). Elsevier Pub.
18. McDowell, C. L. M., & Forrest, E., & Bogdan, (2018). *Impact of AI and Virtual Personal Assistants on Marketing*, 9. Copyright 2018, Elsevier Pub.
19. Syam, N., & Sharma, A., (2018). Waiting for A sales renaissance in the 4th industrial revolution: Machine learning and AI in sales research and practice. *Industrial Mktg. Mgmt., 69*, 135–146. Elsevier Pub.
20. Siau, K., & Yang, Y., (2017). *Impact of AI, Robotics and Machine Learning on Sales and Marketing*. AI, Deep Learning and Machine learning on Marketing.
21. Gursay, D., (2018). Future of hospitality marketing and management research. *Tourism Mgmt. Perspective, 25*, 185–188. Elsevier Pub.
22. Michaels, V., (2018). *How Will AI Impact Digital Marketing in 2018?* TNW (Contributors).
23. Ema, A., Hattori, H., Osawa, H., & Akiya, N., (2015). *Ethics and Social Responsibility: Case Study of a Journal Cover Design Under Fire*. ACM Digital Library, CHI2015, Crossings, Seoul, Korea.
24. Durbach, I. N., & Hofmeyr, J. H., (2007). Interactions between Market Barriers and Communication Networks in Marketing Systems. *The 6th Intl. Joint Conf. on Autonomous Agents and Multi-Agent Systems*.
25. Conick, H., (2016). *The Past, Present, and Future of AI in Marketing*. Marketing News, Publications, American Marketing Association.
26. Goel, R., Sahai, S., et al., (2017). An empirical study to enquire the effectiveness of digital marketing in the challenging age with reference to the Indian economy. *Pertanika Journal of Social Sciences and Humanities Open Access, 25*(4), 1569–1584.
27. Goel, R., Sahai, S., Vinaik, A., & Garg, V., (2019). Moving from cash to cashless economy: A study of consumer perception towards digital transactions. *International Journal of Recent Technology and Engineering, 8*(1), 1220–1226. Retrieved from: www.scopus.com (accessed on 1 December 2020).

28. Malik, P., Singh, G., Sahai, S., Bajpai, C., Goel, R., & Krishnan, C., (2017). Consumer awareness of digital payment with special reference to the village area. *Pertanika Journal of Social Sciences and Humanities, 25*(4), 1585–1600. Retrieved from: www.scopus.com (accessed on 1 December 2020).
29. Sahai, S., Goel, R., Garg, V., & Vinaik, A., (2019). Impact of digitization on impulse buying–what makes the customer bite the bait. *International Journal of Innovative Technology and Exploring Engineering, 8*(7), 2948–2952. Retrieved from: www.scopus.com (accessed on 1 December 2020).
30. Sahai, S., Goel, R., Venaik, A., & Garg, V., (2019). Impact of digital commerce on fashion industry to gain customer loyalty. *International Journal of Engineering and Advanced Technology, 8*(5), 730–740. Retrieved from: www.scopus.com (accessed on 1 December 2020).
31. Vinaik, A., Goel, R., Sahai, S., & Garg, V., (2019). The study of interest of consumers in mobile food ordering apps. *International Journal of Recent Technology and Engineering, 8*(1), 3424–3429.

INDEX

B

C

P

Q

R

S